167
Structure and Bonding

Aims and Scope

The series *Structure and Bonding* publishes critical reviews on topics of research concerned with chemical structure and bonding. The scope of the series spans the entire Periodic Table and addresses structure and bonding issues associated with all of the elements. It also focuses attention on new and developing areas of modern structural and theoretical chemistry such as nanostructures, molecular electronics, designed molecular solids, surfaces, metal clusters and supramolecular structures. Physical and spectroscopic techniques used to determine, examine and model structures fall within the purview of *Structure and Bonding* to the extent that the focus is on the scientific results obtained and not on specialist information concerning the techniques themselves. Issues associated with the development of bonding models and generalizations that illuminate the reactivity pathways and rates of chemical processes are also relevant

The individual volumes in the series are thematic. The goal of each volume is to give the reader, whether at a university or in industry, a comprehensive overview of an area where new insights are emerging that are of interest to a larger scientific audience. Thus each review within the volume critically surveys one aspect of that topic and places it within the context of the volume as a whole. The most significant developments of the last 5 to 10 years should be presented using selected examples to illustrate the principles discussed. A description of the physical basis of the experimental techniques that have been used to provide the primary data may also be appropriate, if it has not been covered in detail elsewhere. The coverage need not be exhaustive in data, but should rather be conceptual, concentrating on the new principles being developed that will allow the reader, who is not a specialist in the area covered, to understand the data presented. Discussion of possible future research directions in the area is welcomed.

Review articles for the individual volumes are invited by the volume editors.

In references *Structure and Bonding* is abbreviated *Struct Bond* and is cited as a journal.

More information about this series at http://www.springer.com/series/430

Stuart A. Macgregor • Odile Eisenstein
Editors

Computational Studies in Organometallic Chemistry

With contributions by

N.M.S. Almeida · F.M. Bickelhaupt · K.J.T. Carr ·
O. Eisenstein · D.H. Ess · S.J. Gustafson · J. Jover ·
C.R. King · Z. Lin · A. Lledós · S.A. Macgregor · F. Maseras ·
D. McKay · R.G. McKinlay · C.L. McMullin ·
M.J. Paterson · L. Perrin · G. Ujaque · P. Vidossich ·
L.P. Wolters

 Springer

Editors
Stuart A. Macgregor
Institute of Chemical Sciences
Heriot-Watt Unversity
Edinburgh
United Kingdom

Odile Eisenstein
Institut Charles Gerhardt
UMR 5253-CNRS-UM-ENSCM
Université de Montpellier
Place E. Bataillon
Montpellier
France

ISSN 0081-5993   ISSN 1616-8550   (electronic)
Structure and Bonding
ISBN 978-3-319-31636-9   ISBN 978-3-319-31638-3   (eBook)
DOI 10.1007/978-3-319-31638-3

Library of Congress Control Number: 2016939982

Preface

In compiling this volume *Computational Studies in Organometallic Chemistry*, we have invited seven contributions that showcase how state-of-the-art quantum methods can address practical problems in molecular inorganic chemistry and provide insights into the factors that underpin the behaviour of these systems. In the first chapter Perrin, Carr, McKay, McMullin, Macgregor and Eisenstein present a description of current practices for the exploration of reaction mechanisms. This is further illustrated by Lin's chapter on the structure and reactivity of boryl complexes. Jover and Maseras then describe the use of QM/MM methods in modelling selectivity, and Vidossich, Lledós and Ujaque detail how molecular dynamic simulations have been applied in this area. Almeida, McKinlay and Paterson describe what is now possible in modelling excited states and associated reactivity. Tools to understand the origin of energy barriers are the focus of the chapters presented by Wolters and Bickelhaupt who outline the energy decomposition analysis (EDA) approach, while King, Gustafon and Ess apply this technique to the topic of C–H activation.

As well as thanking all the contributing authors, the editors would like to dedicate this volume to the memory of Tom Ziegler, a true pioneer in the development of methods and their applications in inorganic chemistry.

Stockholm, Sweden
4 September 2015

Stuart A. Macgregor
Odile Eisenstein

Contents

Struct Bond (2016) 167: 1–38
DOI: 10.1007/430_2015_176
© Springer International Publishing Switzerland 2015
Published online: 11 September 2015

Modelling and Rationalizing Organometallic Chemistry with Computation: Where Are We?

Lionel Perrin, Kevin J.T. Carr, David McKay, Claire L. McMullin, Stuart A. Macgregor, and Odile Eisenstein

Abstract In this chapter, a perspective on how the field of applied computational organometallic chemistry has developed since the mid-1980s is presented. We describe the way in which the modelling of chemical systems has evolved over time, using metallocene chemistry as an example, and highlight the successes and limitations of simple models that were mandatory in the early days of the discipline. A number of more recent case studies are then presented where the full experimental system is now employed and a more quantitative outcome is sought. This includes examples from the Ce-mediated hydrogenation of pyridine, Rh-catalysed C–H bond activation and functionalization, Pd-catalysed azidocarbonylation and phenyl iodide activation at Ru(II) complexes. We conclude with our take on the title question.

Keywords Bond activation · Computational chemistry · DFT · Dispersion correction · Mechanism · Modelling · Organometallics · Selectivity

L. Perrin (✉)
Université Claude Bernard Lyon 1, CNRS UMR 5246, Institut de Chimie et Biochimie Moléculaires et Supramoléculaires, CPE Lyon, 43 Boulevard du 11 Novembre 1918, 69622 Villeurbanne cedex, France
e-mail: lionel.perrin@univ-lyon1.fr

K.J.T. Carr, C.L. McMullin, and S.A. Macgregor (✉)
Institute of Chemical Sciences, Heriot-Watt University, Edinburgh EH14 4AS, UK
e-mail: S.A.Macgregor@hw.ac.uk

D. McKay
Present address; School of Chemistry, University of St. Andrews, North Haugh, St. Andrews KY16 9ST, UK

Institute of Chemical Sciences, Heriot-Watt University, Edinburgh EH14 4AS, UK

O. Eisenstein (✉)
Institut Charles Gerhardt, CNRS UMR 5253, Université de Montpellier, Place E. Bataillon, 34095 Montpellier, France
e-mail: odile.eisenstein@univ-montp2.fr

Contents

1 Introduction

Computational chemistry has made a sustained contribution to the understanding of chemical reactions. In 1965, three landmark communications in the *Journal of the American Chemical Society* from Woodward and Hoffmann marked a paradigm shift in the way that chemists approached the problem of mechanism and reactivity. These three papers which dealt with, in turn, electrocyclic reactions [1], cycloadditions [2] and sigmatropic rearrangements [3] laid down the fundamental bases for the theoretical treatment of concerted reactions and culminated in the formulation of the Woodward–Hoffmann rules [4]. This way of linking experiment and theory opened the door for the application of qualitative methods by Fukui, Hoffmann and others in the areas of organic, inorganic and then organometallic chemistry. In the years up to the award of the Nobel Prize, shared with Fukui, in 1981 [5], Hoffmann developed the concept of the isolobal analogy that highlights the commonalities that exist right across these fields [6]. Running in parallel throughout this period were the developments in theoretical chemistry championed by Pople and Kohn that were ultimately recognized with the award of the Nobel Prize in 1998. The qualitative approach of Hoffmann and Fukui emphasized the underlying electronic structure of functional groups and often exploited symmetry and topological arguments to account for patterns in structure and reactivity. At the other extreme, both wave function theory and density functional theory (DFT) offered the promise of quantitative accuracy.

In recent decades, computational chemistry as applied to organometallic chemistry has further built on the ideas of Hoffmann and has increasingly looked to exploit the sophisticated tools developed and made available by theoretical chemists [7, 8]. It seems that we are now at the point where computational chemistry can provide a quantitative representation of chemical reactivity. This, coupled to the fundamental underlying principles derived from earlier work and modern techniques of electronic structure analysis, is seemingly leading us towards the "Holy Grail" of predicting reactivity in a quantitative way that may truly inform the design of experimental systems.

In this chapter, we present our perspective on how the field of applied computational organometallic chemistry has developed since the mid-1980s. Unless otherwise stated, all the studies that we present hereafter were carried out at the DFT level using a variety of functionals. Although we do not discuss computational methods as such, we will mention them in some cases where the choice of functional proved important. Our citations to computational methods are thus very limited and specific. In Sect. 2, we describe how the modelling of chemical systems evolved and how even simple models are capable of providing fundamental insights. Studies on metallocene chemistry are used to illustrate this evolution. In Sect. 3, we move to the present day where the full experimental system is now routinely employed and a more quantitative outcome is sought. This approach is illustrated with a number of case studies emerging from our recent work. We then give our opinion regarding the title question.

2 The Development of Chemical Models

Setting the right chemical model is essential to the computational study of chemical behaviour. This applies to problems that are associated with all aspects of chemistry, i.e. structure, properties and reactivity patterns. In this chapter, we focus on structural and reactivity issues. Physical properties, for instance, the calculation of thermodynamic parameters (formation energies, pK_a, etc.) and spectroscopic properties (vibrational, electronic, NMR, etc.), present different challenges that will not be documented here.

For the vast majority of cases, a chemical reaction occurs between compounds in solution, often referred to as wet chemistry, and this is our primary concern in this chapter. This excludes the modelling of atmospheric chemistry or reactions taking place in laser beams, mass spectrometers and related experimental conditions. We also limit this chapter to molecular chemistry, and we therefore do not discuss the modelling of surface and supported chemistry or that occurring within solids. Representing experimental wet chemistry is in fact very complex, as many (or most) chemical transformations involve a large number of components, present in various concentrations, which participate in all or part of the multistep sequences that characterize most reactions. Ideally, the chemical model will capture all these aspects, informed by extensive chemical knowledge and supported by physical studies (kinetics, characterization, etc.). In many cases, this can be done; however, surprises exist, and unexpected participants, sometimes even established by computation, may play a role [9]. Another problem is that the nature of the reactants and products should be clearly identified, and this can be particularly challenging in catalysis. For instance, this is often the case in Pd chemistry where the exact nature of the coordination sphere around the metal is not always clearly established. Usually, the products of interest are assigned, but one should be aware of (and eventually concerned about) possible side reactions. Such loose ends can be extremely informative [10, 11]. Bob Crabtree encapsulated these points when he wrote "Computational mechanistic work is also playing a larger role in guiding catalyst research. With the wrong model for the active catalyst, computation is at least severely hampered, if not doomed to failure" [12]. To paraphrase: "rubbish in-rubbish out". It is however very refreshing that calculations can sometimes correct the identity of the products following an erroneous assignment from experiment, as shown by some text book cases [13].

Molecules are composed of functional groups and structural components, where the latter are generally considered as spectators. It is thus tempting to replace these "spectators" by a simplified model that represents the real group. This was done extensively in early work and allowed theoretical chemists to initiate computational studies of reaction mechanisms when computational power and available methodologies placed practical constraints on what could be tackled. Nevertheless, valid results were obtained with such simplified models, and while acknowledging some limitations, the subsequent study of these with more sophisticated models did not necessarily resolve all the issues. The process by which a model is selected is therefore also of interest, and understanding the limitations of a model can provide valuable insight.

2.1 The Case of Metallocene Derivatives

In this section, we draw on examples from the rich stoichiometric and catalytic chemistry of metallocenes, an area where computational chemistry has played a role since the mid-1980s. This provides an ideal opportunity to show how the evolution of models and the availability of different methodologies have impacted on the field.

2.1.1 The First Models

We consider here the modelling of metallocene derivatives of the type "Cp"$_2$MX$_n$ where "Cp" is a cyclopentadienyl derivative, C_5R_5 ($Cp = C_5H_5$), M is a transition metal or a group 3 or lanthanide element and X_n are n X-type ligands. In "Cp"$_2$MX$_n$, "Cp" is an anionic ligand and in early work was modelled by a monoatomic ligand. The angle "Cp"$_{cent}$–M–"Cp"$_{cent}$, where "Cp"$_{cent}$ is the centroid of the "Cp" ring, is often around 140° especially when M is a transition metal or $4f$ element. The basic molecular orbital pattern of Cp_2MX_n is properly reproduced by extended Hückel theory (EHT) calculations, where this computationally inexpensive approach allowed "Cp" to be C_5H_5 itself [14]. The $3d$ orbitals of Cp_2TiH_2 are thus split into a pattern that is characteristic of a tetrahedral ligand field. For this reason, Cp_2TiH_2 was even represented by TiH_4 in early ab initio calculations [15]. Today, it could seem rather rash, to say the least, to use H as a model for Cp, but this model succeeded because it had d orbitals of similar shape and with similar energy splitting. However, a hydride is a two-electron donor, while a cyclopentadienide anion is a six-electron donor. Therefore, in subsequent work, a Cl_2TiH_2 model was preferred. First, the Cl–Ti–Cl angle reproduced the 140° angle more accurately, possibly because of the repulsion between the Cl lone pairs. Furthermore, Cp and Cl are isolobal, each having a total of six electrons in their occupied a- and e-type symmetry orbitals that

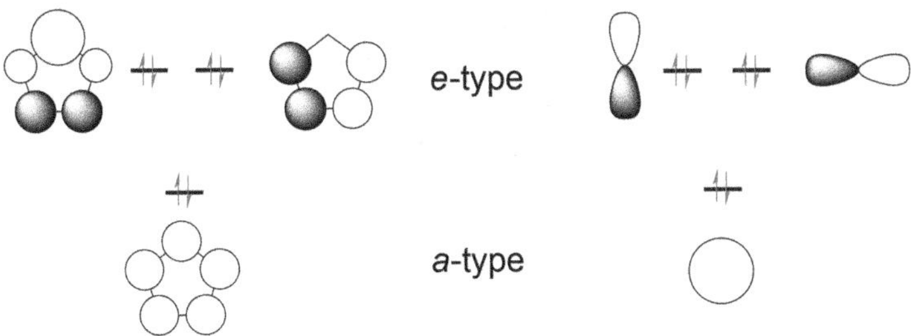

Scheme 1 The occupied π orbitals of cyclopentadienyl and p orbitals of chloride anions with approximate local symmetry labels

interact with the metal d orbitals in the same qualitative manner (Scheme 1). However, the isolobal analogy is a "qualitative" and not a "quantitative" concept [16]. Therefore, the different spatial characters and energies of the π orbitals of Cp and the lone pairs of Cl result in quantitatively different M–Cp and M–Cl interactions. While in early work this was the only computationally affordable model, it still led to very insightful results. One of the earliest studies was on σ-bond metathesis by Steigerwald and Goddard [17] who used general valence bond (GVB) calculations to determine the H/H exchange. Associated analysis showed that there is no longer symmetry control in the σ-bond metathesis (also called the $[2\sigma^2 + 2\sigma^2]$ reaction) because of the presence of the d orbitals on the metal centre. In contrast, in organic chemistry, such a $[2\sigma^2 + 2\sigma^2]$ reaction is symmetry forbidden. The very low computed barrier (ca. 2 kcal/mol for Cl_2TiH^+) could not be compared to any experimental data but was consistent with observations of rapid H/D exchange. This seminal study demonstrated that the essential electronic properties of the $\{Cp_2M\}$ metallocene fragments were captured by the crude $\{Cl_2M\}$ model. This study also indicated that the symmetry rules established for organic systems do not apply in transition metal chemistry, and this is still accepted today.

We will now present a number of studies on Group 3, Group 4 and lanthanide complexes. In all cases, an effective core potential (ECP) is used to represent the inner electrons of atoms. This choice is straightforward for transition metals, but some comment is required for lanthanides. These have partially filled $4f$ shells that, in principle, require an explicit treatment (with a so-called small core ECP), which would considerably increase the computational effort. However, as calculations showed that $4f$ orbitals do not participate significantly in bonding [18], it was tempting to use a large core ECP which places the $4f$ electrons in the core. Test studies comparing small and large core ECPs showed that structural features are properly reproduced, with the latter giving metal–ligand distances that are only slightly too long. The large core also gave acceptable energy profiles compared to those previously reported by Goddard et al. for H/H exchange at Cl_2MH [16]. A large core ECP has therefore been employed in a large number of studies of the reactivity of lanthanide complexes since 2000. It should be noted, however, that each large core ECP is specific to a given oxidation state of the metal, and this is a limitation for the study of redox reactions [19]. In earlier studies, the influence of the lanthanide contraction on reactivity was shown to be small for the test reactions of X_2MH with H_2, CH_4 and SiH_4 [20–22]. Therefore, conclusions derived from a given lanthanide should be applicable to all the other lanthanides in an equivalent redox state.

2.1.2 Refining Simple Models

Initial studies tested different ways to model the Cp ligand by comparing energy barriers for the H/H exchange reaction of X_2MH with H_2, where $M = Ln^{III}$ and $X = H$, Cl, Cp, or an effective group potential, EGP [16, 23, 24]. The purpose of an EGP is to represent only the valence orbitals of a chemical group, here the Cp anion, at a low computational cost. While this approach reproduced the data for the "learned" system well, it has not been widely used due to the rapid increase in computational power.

For the H/H exchange reaction, the computed barriers depend upon the nature of X, with those for $X = Cp$ (and EGP) being consistently lower than those for $X = Cl$. One reason for this discrepancy is that Cl is a weaker electron donor than Cp, likely due to both its high electronegativity and smaller overlap with the metal orbitals. Due to the electron attractor power of the chloride ligands, there is no sufficient hydridic character on the hydrogen atom. This is crucial as the σ-bond metathesis transition state is a four-centre ring in which the central hydrogen moves as a proton between two hydrides (Scheme 2). Low hydridic character on the α-hydrogen atoms therefore raises the energy barrier [20].

This analysis is consistent with some earlier studies where difficulties were encountered in representing Cp with Cl. In a study of Cp_2MH_3 ($M = Nb$, Ta), the use of Cl for Cp overstabilizes the dihydrogen/hydride isomeric form over the observed trihydride form [25, 26]. Similar outcomes were obtained when modelling the reactions of Cp_2MH ($M = Zr^+$, Sc) with acetylene [27]. Overall, these studies suggested that the Cl for Cp model might have difficulties when dealing with subtle energy differences.

An example where the Cl for Cp model was extremely successful in providing fundamental insight was in modelling the pattern of the σ-bond metathesis reactions shown in Scheme 3. Regardless of the model used for the metal fragment, the reactions shown in Eqs. 1a and 1c had very high computed barriers, and indeed, the equivalent reactions with Cp_2MH have never been experimentally observed. In these two processes, the four-centre transition state has a CH_3 group at the β-position (Scheme 3a, c), and this is energetically unfavourable as it forms a local hypervalent CH_5^- moiety. In contrast, when H is at the β-position, low energy

Scheme 2 Transition state for $[2\sigma^2 + 2\sigma^2]$ σ-bond metathesis between X_2LnH (denoted [Ln]H) and H_2

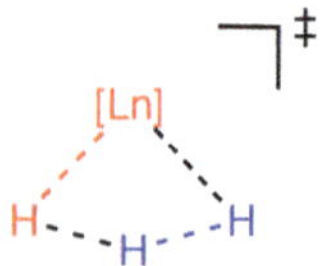

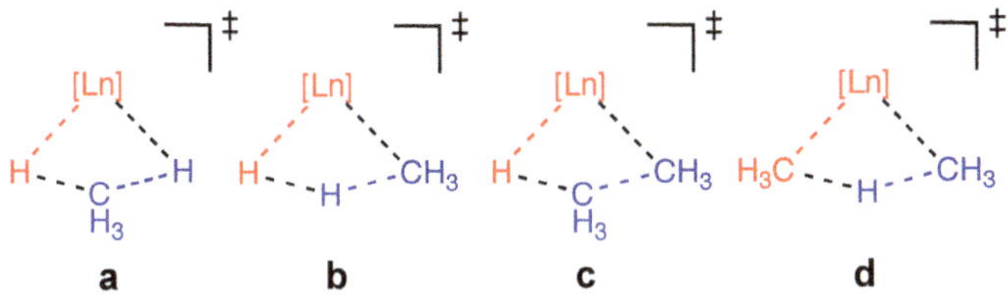

Scheme 3 Transition states for (**a**) Eq. 1a, (**b**) Eq. 1b, (**c**) Eq. 1c and (**d**) Eq. 1d for $M = d^0$ metal or Ln^{III}, L_2LnR is denoted [Ln]R

barriers are computed, and again, this is independent of the model selected for the metal fragment. Note that Eqs. 1b and 1c are exothermic for hydride formation [21, 23, 28]. The low energy barrier to reach the transition state in Scheme 3d rationalizes the remarkable observation that $(C_5Me_5)_2LuCH_3$ activates CH_4 [29].

$$Cl_2M - H + H - CH_3 \xrightarrow{\quad\times\quad} Cl_2M - H + H - CH_3 (\Delta E = 0) \tag{1a}$$

$$Cl_2M - H + H - CH_3 \rightarrow Cl_2M - CH_3 + H - H (\Delta E > 0) \tag{1b}$$

$$Cl_2M - CH_3 + H - CH_3 \xrightarrow{\quad\times\quad} Cl_2M - H + CH_3 - CH_3 (\Delta E > 0) \tag{1c}$$

$$Cl_2M - CH_3 + CH_3 - H \rightarrow Cl_2M - CH_3 + H - CH_3 (\Delta E = 0) \tag{1d}$$

2.1.3 Towards Experimental Complexity

While modelling Cp with Cl has provided useful insights, more quantitative aspects suggested that this approach was reaching its limit. In 2002, a combined experimental and computational study highlighted the intricate influence of the Cp substituents in zirconocene derivatives [30]. Experimentally, CO stretching frequencies, reduction potentials of dichloro complexes and UV and NMR data were compared with computed electronic affinities and CO stretching frequencies. This study showed that the Cp substituents influence not only the electron density on the metal but also the Cp_{cent}–M–Cp_{cent} angle, and both these effects have important consequences on reactivity. This last aspect is particularly important when the two rings are connected as in *ansa* complexes.

Nevertheless, at that time there was still a strong incentive for using the simplest C_5H_5 ring since this was the only practical way to perform reactivity studies. Indeed, the ten methyl groups in $(C_5Me_5)_2M$ not only raise the number of electrons and basis functions compared to $(C_5H_5)_2M$ but also the number of degrees of freedom to be properly optimized in order to avoid spurious rotations of the methyl groups. With larger ligands, as in $[1,2,4-(Me_3C)_3C_5H_2]_2M$ derivatives, the difficulty becomes even worse as the rotation of the ring, that of the substituents and the angle between the rings are probably all coupled (see Sect. 3.1). Selecting C_5H_5 as a universal representation of any cyclopentadienyl ring looked indeed a wise choice. As a universal model, it was successful...until it failed! In the following, we

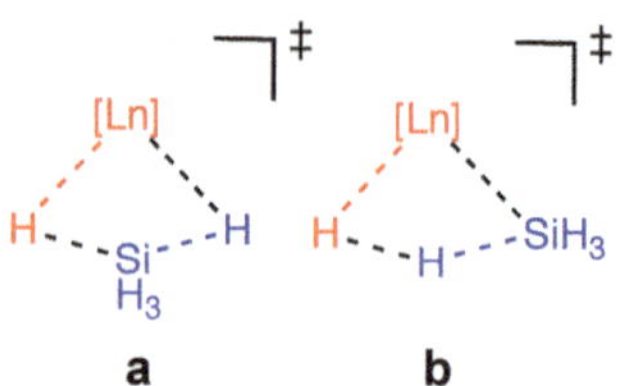

Scheme 4 Four-centre transition states in the reaction of X_2MH with SiH_4 in which the SiH_3 group is at the β- or α-position, labelled **a** and **b**, respectively

describe the rise and decline of this model by highlighting the reasons for its successes and failures.

Studies of the reaction of Cp_2MR (R = H, CH_3) with H_2 and CH_4 were in qualitative agreement with the previous work using the Cl_2MR model as presented in Scheme 3 [20, 23, 31, 32]. The essential result, i.e. the high energies of the transition states 3a and 3c, is retained; moreover, it can be easily understood why the reactions of the same X_2MR species with silanes show a very different energy profile [20]. Thus, the four-centre transition states with the silyl group at either the α- or the β-position (Scheme 4) are both accessible and have similar energies for any d^0 metal (Sc, Y or La, Lu and Sm), any model of ligand (X = Cl, Cp) [33] and any silane (SiH_4, $PhSiH_3$, CH_3SiH_3) [22, 34, 35]. This also accounted for the dehydropolymerization of silanes catalysed by Cp_2M derivatives and modelled by Cl_2MH (M = Sc, Y, La, Lu, Sm) [33]. Thus, the fundamental difference between σ-bond metathesis reactions involving alkanes and silanes is properly mimicked by calculations *at any level of modelling*, and this reflects the different propensity of the alkyl and silyl groups to become hypervalent.

Computational studies of the reactions of metallocene derivatives with various heterosubstituted alkanes and arenes were also carried out, where the very bulky $1,2,4\text{-}(Me_3C)_3C_5H_2$ (denoted Cp′) ligand was modelled by C_5H_5. In retrospect, this simplification looks drastic, but the first results were encouraging and did suggest that the six ^tBu groups on the cyclopentadienyl rings could be neglected.

$$Cp'_2CeH + CH_nF_{4-n} \rightarrow Cp'_2CeF + CH_{n+1}F_{3-n}(n = 1, 3) \tag{2a}$$

$$Cp'_2CeH + CH_3X \rightarrow Cp'_2CeX + CH_4(X = Cl, Br, I, NMe_2) \tag{2b}$$

$$Cp'_2CeH + ROR \rightarrow Cp'_2CeOR + RH \ (R = alkyl) \tag{2c}$$

As shown in Eqs. 2a [36, 37], 2b [38] and 2c [38–42], the reactions of Cp'_2CeH with simple organic molecular systems of the type RX (or ROR) form Cp'_2CeX (or Cp'_2CeOR) and RH. One possible mechanism that could account for this is a σ-bond metathesis via a four-centre transition state with CH_3 at the β-position, perhaps driven by the formation of the strong Ce–X (or OR) bond (**TS₀**, Fig. 1). However, this was found to have a very high computed barrier for any X. In fact, the preferred reaction pathway is a two-step process, which starts by deprotonation of the Ce-coordinated CH_3X by the hydride via **TS₁** to yield a carbenoid intermediate, **CARB** (subsequently isolated for OR = OMe, Scheme 5) [38], and H_2. The second step via **TS₂** is a transfer of the methylene group from **CARB** to a single bond such as H_2 or C–H. This sequence, which can be regarded as an α-elimination followed

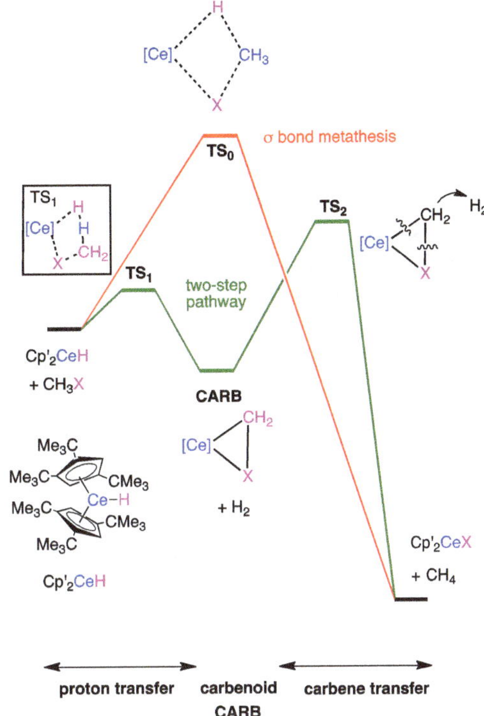

Fig. 1 Direct σ-bond metathesis vs. carbenoid two-step mechanism for formal σ-bond exchange between [Ce]–H and C–X

Scheme 5 Carbenoid complex (**CARB**) represented with experimental Cp′ ligand

by a quenching of the carbene by a nearby σ-bond, is reminiscent of the Simmons–Smith reaction.

The success in representing these reactions and, moreover, proposing a pathway that was novel for lanthanides was highly encouraging. However, not all aspects of the experimental observations could be reproduced. A pathway for C–F activation in C_6F_6 and C_6F_5H [43] was proposed with the intermediate formation of a benzyne observed by quenching experiments. However, the calculations were never able to reproduce the observation that CH and CF activation of partially fluorinated arenes had similar rates [44, 45]. This discrepancy was thought to be related to the difficulty of quantitatively representing the activation of the very strong C–F bond in which electron correlation could be important.

Further indications of difficulties in representing experimental observations came with the reaction of Cp'_2CeCH_2Ph with CH_3X (X = F, Cl, Br, I) in which an intact CH_3 group is transferred to the benzyl ligand to form Cp'_2CeX and ethylbenzene [40]. While labelling studies excluded proton transfer as an initial reaction step, calculations found this process to be kinetically competitive with an S_N2-like C–C bond formation between the benzyl and CH_3F with concerted C–F bond cleavage. This latter transition state is stabilized by interaction between Ce and the π density of the aryl group. Furthermore, the benzyl complex was found to adopt two different structures in the solid state, only one of which could be reproduced by calculations in the gas phase. It was assumed that the other structure was the result of crystal packing [46].

A study of methane metathesis at X_2MCH_3 (M = Sc, Y, Lu) by Cramer et al., where X = H, Cl, C_5H_5 (Cp) and C_5Me_5 (Cp*), highlighted the need for the proper modelling of metallocene derivatives, especially when comparing competitive pathways [47]. Dimerization, unimolecular methane ejection (with formation of a cyclometallated "tuck-in" complex) and bimolecular methane metathesis were compared (Scheme 6). Experiments by Watson et al. had suggested that the last route was preferred for M = Y and Lu, while the unimolecular reaction was operative for Sc. Not surprisingly, the calculations showed that results with X = H and Cl were significantly different from those with X = Cp*, while the model for X = Cp performed better but still gives underestimated energy barriers for all pathways. Even more importantly, to quote the authors, "Replacing Cp* by Cp not only affects the absolute activation enthalpies by about 3 kcal/mol but also degrades the quality of the relative activation enthalpies". Comparing the pathways for all X and M revealed that methane loss to give the "tuck-in" complex is competitive with the other pathways for Sc and actually becomes preferred with the Cp* ligand. Later on, methane metathesis at metallocene–propyl complexes was studied as a step in the hydromethylation of propene, leading to a study of alkyl group substituent effects. Using the MPW1K functional, which the authors had shown to reproduce thermodynamic data well, it was confirmed that the reaction is bimolecular for Lu and that large alkyl ligands and small ionic radii (Sc vs. Lu) make the uni- and bimolecular reactions competitive [48]. Related studies of the full reaction pathway of hydromethylation or hydrosilylation of propene, including

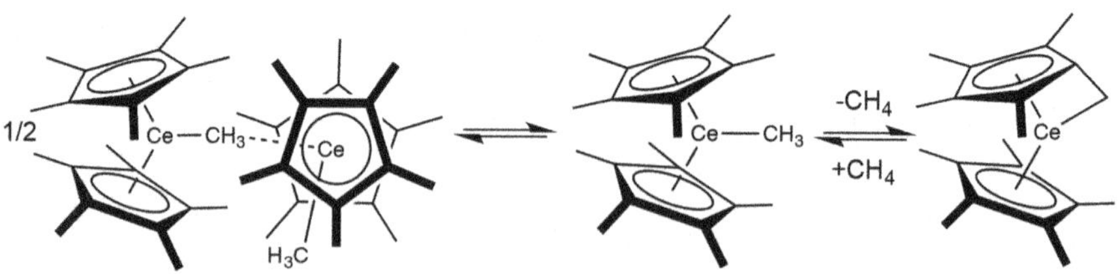

Scheme 6 Inter- and intramolecular reactivity of the Cp_2*CeCH_3 complex

pathways for the formation of side products, confirm the strong influence of the ligand bulk and the metal ionic radius on the reaction mechanism. The steric and electrostatic effects need thus to be properly represented in the models chosen to model the reactive systems [49–52].

The simple Cp-based model for substituted metallocene derivatives was still used successfully for rationalizing the reaction of Cp'_2CeH with CH_3OSiMe_3 to give a methoxide product (Eq. 3a; where Cp'_2CeH, denoted $[Ce]'H$, is modelled as Cp_2CeH, Scheme 7) [39]. However, this system required comparison with the related "tuck-in" metallacycle, denoted $\{Ce\}'CH_2$, which gave different products (Eq. 3b, modelled as $\{Ce\}CH_2$, Scheme 7). With CH_3OSiMe_3, calculations indicate the reaction with Cp_2CeH proceeds in a single σ-bond metathesis step involving the Ce–H and O–SiMe$_3$ bonds, with SiMe$_3$ at the β-position of the four-centre transition state (cf. Scheme 4a). In contrast, the preferred reaction pathway for the tuck-in $\{Ce\}CH_2$ complex is a multistep process initiated by deprotonation of the ether methyl group, followed by migration of the SiMe$_3$ from the oxygen atom onto the newly formed CH_2 moiety, a process known in organosilicon chemistry under the names of the retro-Wittig, retro-Brook or Wright–West rearrangement (Eq. 3b). Thus, the proton transfer associated with the methyl C–H bond activation (α-CH in Fig. 2) is preferred over the SiMe$_3$ migration (*metathesis SiMe₃* in Fig. 2) with the

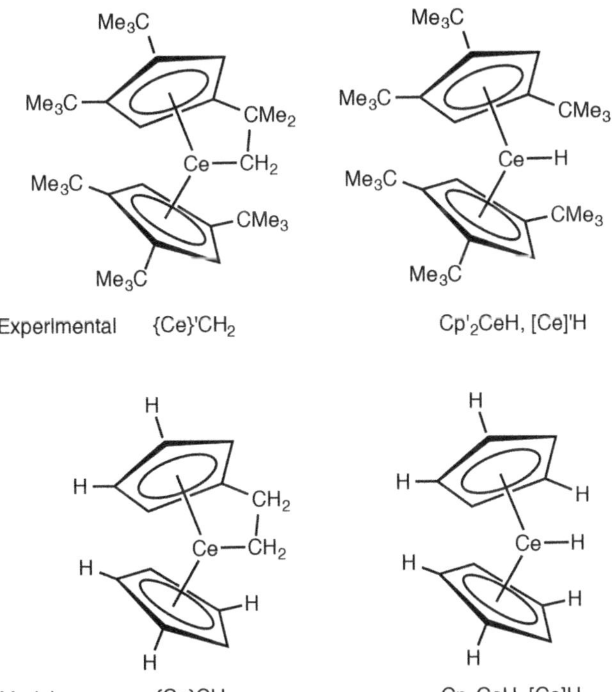

Scheme 7 The "tuck-in" cerium complex and cerium hydride used in the experiment and models used in the calculations with labelling nomenclature

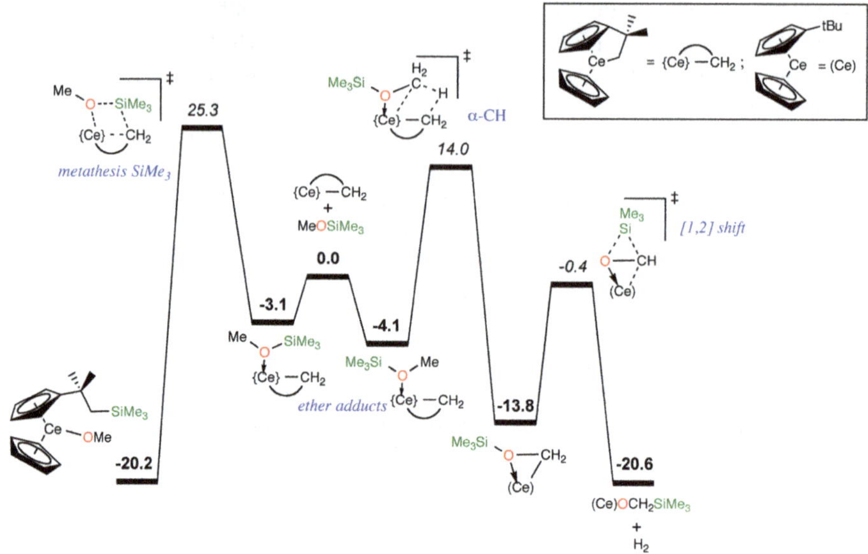

Fig. 2 Energy profile for the reaction of MeOSiMe$_3$ with the tuck-in complex {Ce}$'$CH$_2$ modelled by (C$_5$H$_5$)(C$_5$H$_4$)C(Me)$_2$CH$_2$Ce

tuck-in complex, but neither of these processes occur with the hydride. In retrospect, it was somewhat fortuitous that the computations reproduced the experimental data even in the absence of all the ^{t}Bu groups on the cyclopentadienyl group.

$$Cp'_2CeH + MeOSiMe_3 \rightarrow Cp'_2CeOMe + HSiMe_3 \tag{3a}$$

$$Cp'\left[(Me_3C)_2C_5H_2C(Me)_2CH_2\right]Ce + MeOSiMe_3 \rightarrow Cp'_2CeOCH_2SiMe_3 \tag{3b}$$

A study of the reactions of CH$_3$OSO$_2$CH$_3$ and CH$_3$OSO$_2$CF$_3$ with Cp$'_2$CeH and {Ce}$'$CH$_2$ marked a clear failure of the, up till now, successful and pragmatic modelling of Cp$'$ by C$_5$H$_5$ [41, 42]. These reactions yield a variety of products, all derived from initial C–H activation at either the methoxy or the thiomethoxy groups. The reaction of Cp$'_2$CeH with CH$_3$OSO$_2$CH$_3$ was selective for activation at the thiomethoxy, while both groups reacted at {Ce}$'$CH$_2$ with no selectivity. The reaction between CH$_3$OSO$_2$CF$_3$ with the hydride proceeded by transfer of a methyl cation to form CH$_4$, reminiscent of an S$_N$1 reaction, and promoted by the high stability of the tosylate anion. However, calculations with a Cp$_2$M model were unable to account for these observations, and so larger models in which the Cp$'$ ligand and the tuck-in complex were represented in full were employed. Using this approach, the preference for C–H bond activation in the thiomethoxy group of CH$_3$OSO$_2$CH$_3$ was reproduced. Moreover, with CH$_3$OSO$_2$CF$_3$, the methyl cation transfer became accessible, although alternative pathways initiated by the C–H bond activation remained competitive. In all cases, there was a need to represent

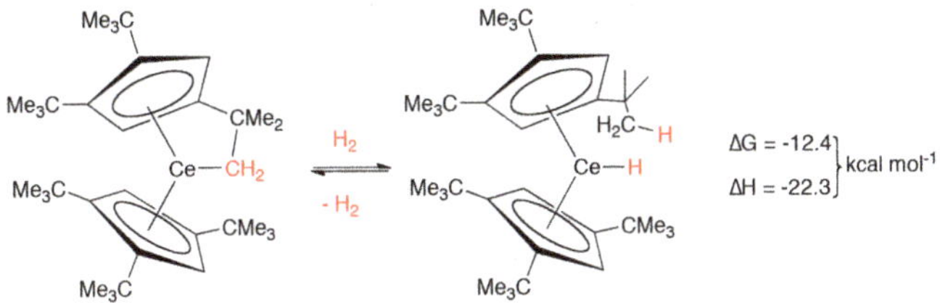

Fig. 3 Equilibrium between the tuck-in $\{Ce\}'CH_2$ complex and $[Ce]'H$ under H_2

properly the energy of hydrogenation of the $\{Ce\}'CH_2$ complex, which can only be properly reproduced with the full model (Fig. 3) [41, 42].

3 Current Practice: Case Studies

Up to this point in the chapter, we have focused on how the level of chemical modelling has developed over the last 30 years, using metallocene systems and the cyclopentadienyl family of ligands in particular as examples. The models employed reflected practical issues such as the availability of computational power and the level of theory that had been implemented in accessible codes. While these placed severe constraints on the nature of the problems that could be addressed, it was nonetheless possible to build a wealth of fundamental (if qualitative) insights into the structure and reactivity of organometallic chemistry.

More recently, greater computer power coupled to the availability of more efficient codes and, in particular, the appearance of DFT as a practical tool have allowed the computational chemist to tackle increasingly complex problems. This is also reflected in greater collaboration between computational chemists and experimentalists; indeed, experimentalists are increasingly adopting computational studies as part of their research programmes. The ability to use "real" (i.e. the actual molecular systems used in experiment) in the calculations, along with developments in methodology (thermodynamic corrections, continuum solvation models and dispersion-corrected functionals), coupled to the routine location of transition states provides access to the modelling of complex, multistep reaction pathways. This can produce a series of computed reaction profiles and catalytic cycles along with a wealth of structural and energetic data. The results are seductive, but to what extent are we at the point where we can rely on our computational results in a quantitative sense?

In the second half of this chapter, we present a series of case studies from recent work in the Montpellier and Heriot-Watt labs. All have been performed in collaboration with experimentalists whose observations have provided distinct challenges to our computational modelling. In most cases, we have managed to arrive at a good understanding of the experiment results. In no case, however, was this straightforward, and we hope that exposing the steps required to achieve consistency will be informative to the reader. Nonetheless, in each project, some inconsistencies between observation and computation persist, and the challenge of marrying the two disciplines remains acute. Such cases should not, however, be viewed as "failures". Theory and computation have often, if not always, been driven by exceptions and the challenging behaviour of pathological systems. To a computational chemist, a "failure" is always of great interest: it provides an opportunity to understand and improve both theory and our understanding of the experiment and to question what the best way is to perform our calculations (or indeed if there is an appropriate calculation that can be performed at all). Calculations can also be used to question experiments, which may then need reconsideration. In one example, the product of reaction of H_2 with a dinuclear Zr dinitrogen complex was first wrongly assigned to a dinitrogen–dihydrogen adduct. DFT (B3LYP) calculations on a system that was very challenging because of its size for the time suggested a structure in which NNH and H were bonded to the two zirconium atoms. Neutron diffraction studies confirmed the structure proposed by the calculations [13]. A constructive tension therefore exists between experiments and theory. Such issues were expounded in a recent combined experimental/theoretical study of the mechanism of alcohol-mediated Moriata–Baylis–Hillman reaction. Plata and Singleton remark on the sensitivity of the computed energy landscape depending on the methodology employed, as well as the importance of experimental results for computational validation. However, they also recognize, despite these difficulties, that calculation provides the only handle on the transition states for the formation of some key intermediates [53, 54].

3.1 Hydrogenation of Pyridine

The following example illustrates how a fully integrated study combining experiment and computation forced a careful re-evaluation of the chemical modelling that ultimately led to the discovery of novel pathways. The room temperature hydrogenation of pyridine to piperidine is mediated by $[1,2,4-(Me_3C)_3C_5H_2]_2CeH$ (denoted [Ce]'H) under H_2 pressure (Fig. 4) [55]. The reaction proceeds via the formation of intermediates **III** and **IV**, both characterized by NMR spectroscopy and X-ray crystallography. Such mild hydrogenation of pyridine is unusual and so prompted questions about the mechanism. Experimentally, kinetic studies proved impractical; however, some intermediates were characterized by X-ray crystallography and/or NMR spectroscopy. This established that the *ortho*-metallated pyridine complex **I**, generated by σ-bond metathesis between [Ce]'–H and the pyridine *ortho*-C–H bonds (Fig. 4), was the entry point to the catalytic cycle.

As with previous work, initial calculations were carried out using Cp_2CeR (denoted [Ce]R). These calculations indicated that H_2 readily adds across the Ce–C bond in **I** to give an *N*-bound pyridine adduct, denoted [Ce]H·pyridine. This transformation is endergonic and proceeds through a low-lying transition state, in good agreement with both the sole observation of **I** and of fast *ortho*-H/D exchange.

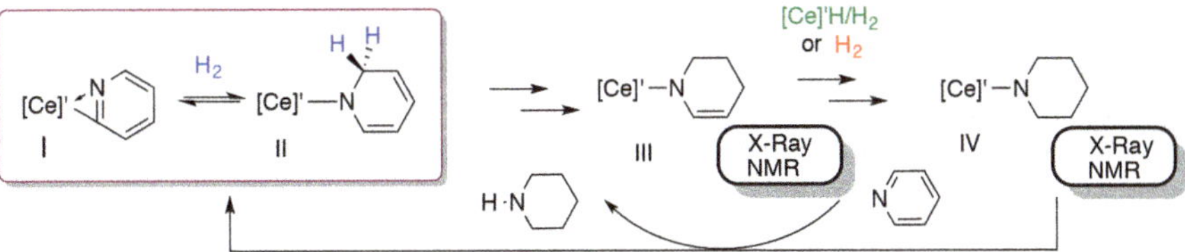

Fig. 4 Catalytic hydrogenation of pyridine by $[1,2,4\text{-}(Me_3C)_3C_5H_2]_2CeH$, $[Ce]'H$, via key intermediate **1**

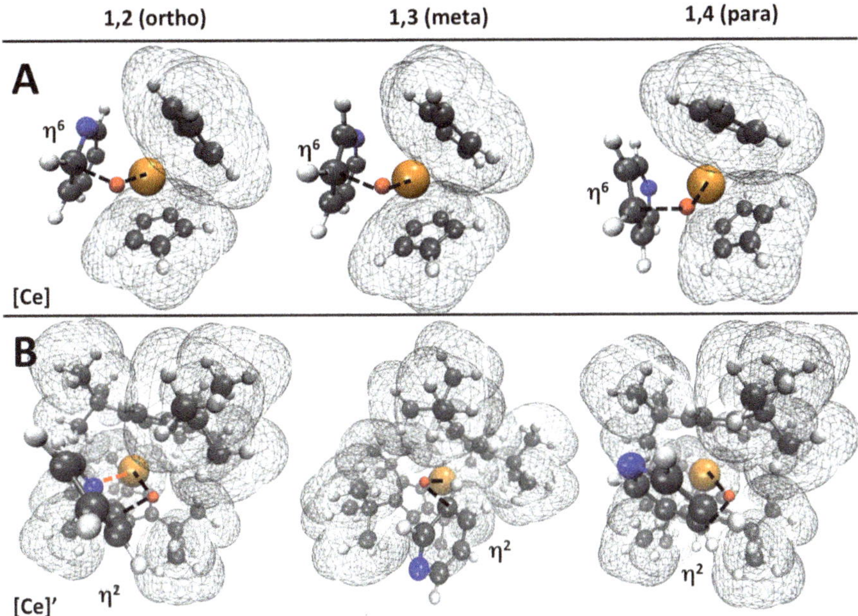

Fig. 5 Space filling diagrams for hydride addition transition states at the *ortho*, *meta* and *para* carbon atoms at models [Ce]H·pyridine (**a**) and [Ce]′H·pyridine (**b**)

From [Ce]H·pyridine, nucleophilic addition of the hydride can occur at either the *ortho*, *meta* or *para* positions of the pyridine, *ortho* attack giving intermediate **II** in Fig. 4. Unfortunately, with the [Ce]R model, the lowest energy transition state corresponded to hydride attack at the *para* position. Precedent for this is seen in yttrium chemistry [56, 57]. However, these hydride addition transition states were all found to have an η^6-bound pyridine (Fig. 5a), retention of which was thought unlikely with six bulky ^{t}Bu groups on the cyclopentadienyl rings (although a related geometry had been seen in $[Ce]'CH_2Ph$ [46]). Regardless of these structural considerations, attack at the *para* position was in contradiction with isotope labelling indicating H/D scrambling *only* at the *ortho* position. Something was wrong.

As suspected, introducing the ^{t}Bu groups into the model does indeed prevent η^6-coordination of pyridine to Ce; instead, the Ce hydride transfer to pyridine in

[Ce]H·pyridine involves an η^2-binding of pyridine and occurs preferably at the *ortho* carbon to form **II**. Thus, a full representation of the Cp′ ligand is needed to correctly model the initial regioselectivity. Following the addition of this first H_2 molecule to the *ortho*-metallated pyridine (**I** to **II**), we were also able to locate pathways for the subsequent hydrogenations. The most accessible of these mono-metallic events involved a heterolytic cleavage of H_2, adding a hydride to the Ce centre and a proton to the *meta* carbon position. This is then followed by hydride transfer to the *para* position to give **III** (Fig. 4). Unfortunately, this hydride transfer step had a barrier of 30 kcal/mol, which is not compatible with the experimental conditions, where the reaction proceeds slowly at room temperature. The high barriers most likely result from the facial interaction between Ce and the pyridyl scaffold, which requires a weakening of the Ce–N interaction.

At this point, since the Cp′$_2$Ce complexes had been represented in full, we were forced to reassess all aspects of the experimental data (an example of "the constructive tension between experiment and theory"!). In fact, the overall rate of the reaction was shown to increase with the Cp′$_2$CeH concentration, suggesting that more than one Cp′$_2$CeH may be involved in the reaction [58]. Bimetallic species therefore had to be considered, but this entailed a plethora of possible conformations (four cyclopentadienyls and a total of twelve ^{t}Bu groups) and so presented a potentially intractable problem. Fortunately, useful information could be derived from the solid-state structures of related dinuclear cerium complexes, which gave a number of possible conformations as reasonable starting points [41, 58]. Initial calculations at the DFT-B3PW91/SMD level again gave very high-energy barriers, most likely because of the numerous close contacts between the bulky groups. Under such circumstances, adding a dispersion correction to account for weak *attractive* interactions is essential, and in the present case, a correction at the D3BJ level [59] led to a stabilization of all bimetallic extrema (minima and transition states) by ca. 20 kcal mol^{-1}. As a result the highest energy transition state among those associated with the tandem hydride addition at the *meta* carbon (Fig. 6) followed by hydrogenolysis of the transient [Ce]′–C$_{para}$ bond, to recover [Ce]′H and to yield **III**, was 8.2 kcal/mol. This reaction sequence is now kinetically accessible and hence compatible with the experimental conditions. The importance of dispersion corrections [60, 61] will be a common theme in all of the case studies discussed here.

Overall, this study lasted several years. In retrospect, the initial use of a small Cp$_2$CeH model for [Ce]′H was totally misleading: this not only gave an incorrect order for the regioselectivity of hydrogenation but would have missed completely the involvement of two metal centres. However, use of the full model of the real system, in itself, did *not* provide the solution. For bulky systems, an appropriate methodology (including dispersion) *must* be used, and an extensive assessment of conformational space is *essential*. In this case, a spread of more than 15 kcal/mol was computed for the key transition state in Fig. 6, reflecting only the different arrangements that can be adopted by the fully substituted cyclopentadienyl ligands. Thus, failing to explore properly the conformational space would have resulted in a misinterpretation of the reaction path.

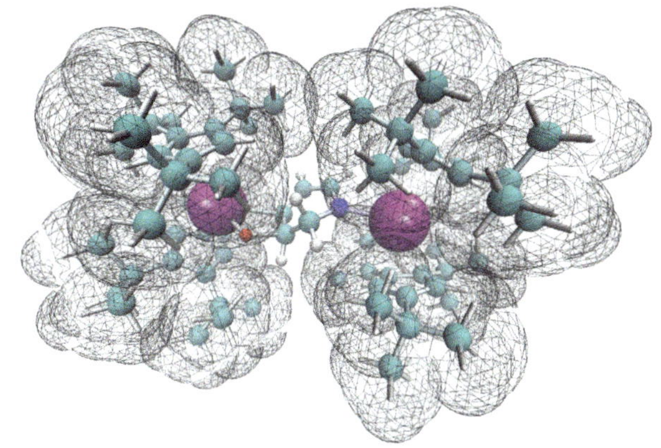

Fig. 6 Structure of the transition state for [Ce]'H addition (*left*) to the *meta* centre of (2*H*-hydropyridyl) [Ce]' (*right*) on the path to generate **III** from **II** Fig. 4

Although a coherent mechanism was eventually defined in this study [55], some issues remained unresolved. In the solid state as well as in solution (according to NMR spectroscopy), the partially or fully hydrogenated pyridine is coordinated to Ce with different Ce–N–C angles (differing by 40° in [Ce]'NC$_5$H$_{10}$ **IV** and [Ce]' NC$_5$H$_{10}$ **III**). This unsymmetrical bonding could not be reproduced, resisting refinement in the method of calculation of the full system (i.e. implementation of dispersion, solvation, choice of functional, increasing basis set size and including small core ECP). In some cases, discrepancies between solid-state and molecular computed structures have been resolved by representing the crystal packing and dispersion corrections [62, 63]. In the present case, both periodic calculations (with and without dispersion corrections) and a molecular model in which the hydrogenated *N*-piperidyl ligand was relaxed inside a cluster of four cerocenes frozen at their solid-state structures were indeed able to reproduce the nonsymmetrical binding of the ligand. However, the distortion observed in solution could not be rationalized.

3.2 C–H Bond Activation and Functionalization at Rh and Ir Half-Sandwich Complexes

3.2.1 Selectivity in C–H Activation

The selectivity of C–H activation is a key issue in developing practical C–H functionalization strategies, as most relevant substrates will have multiple C–H bonds that are available for activation. As part of our ongoing collaboration with the Davies group, a series of competition experiments were devised to probe the relative reactivity of C(sp^2)–H bonds in *N*-alkylimines, **H-L$_{1-5}$**, and phenylpyridine, **H-L$_6$** [64]. These substrates all undergo directed C–H activation in the presence of [Cp*MCl$_2$]$_2$ and NaOAc to give well-defined cyclometallated species as their chloride adducts (Fig. 7). Stoichiometric reactions with **H-L$_3$** in

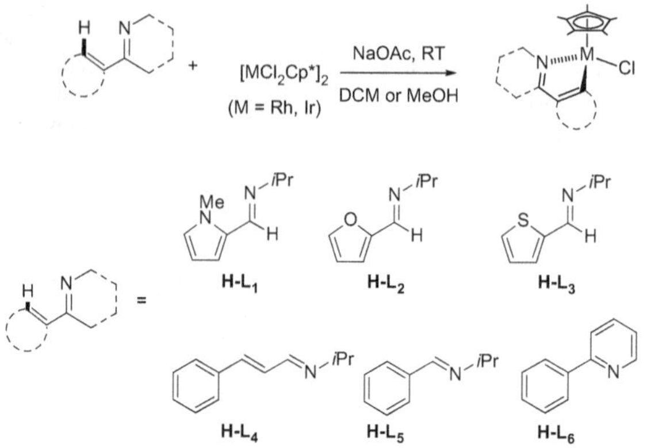

Fig. 7 Acetate-assisted cyclometallation at $[Cp^*MCl_2]_2$ (M = Rh, Ir)

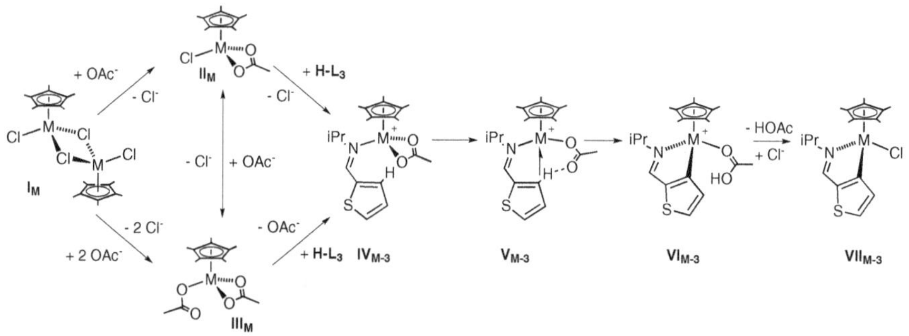

Fig. 8 Mechanism and labelling scheme for the reaction of **H-L$_3$** at $[Cp^*MCl_2]_2$ (**I$_M$**, M = Ir, Rh)

CH_2Cl_2 highlighted the greater efficiency of these cyclometallations at Ir (quantitative, 6 h) compared to Rh (34%, 17 h). H/D exchange experiments in CD_3OD demonstrated irreversible C–H activation at Ir (i.e. kinetic control), but reversible C–H activation at Rh (thermodynamic control).

The mechanism of these reactions is outlined in Fig. 8 and provided two immediate challenges to the computational modelling when compared with the available experimental data. Firstly, the acetate-induced opening of $[Cp^*RhCl_2]_2$, **I$_{Rh}$**, in MeOH is reported to yield an equilibrium mixture of **I$_{Rh}$**, $[Cp^*RhCl(OAc)]$, **II$_{Rh}$** and $[Cp^*Rh(OAc)_2]$, **III$_{Rh}$** [65]. The free energy of these three species must therefore be similar. Secondly, the onward C–H activation to **VII$_M$** is a thermodynamically accessible process for all substrates and at both metals. However, initial test calculations on the reaction of **H-L$_3$** with **I$_{Rh}$** using the BP86 functional and a standard basis set and correcting for MeOH solvent (PCM approach) completely failed to capture these features: the free energies of **II$_{Rh}$** (−25.4 kcal/mol) and **III$_{Rh}$** (−46.5 kcal/mol) were both far too stable relative to **I$_{Rh}$** (set to 0.0 kcal/mol) as well as being very different to each other (Table 1, Entry 1). Moreover, the energy of **III$_{Rh}$** compared to the product, **VII$_{Rh-3}$** (−27.1 kcal/mol), implied that **III$_{Rh}$** would

Table 1 Basis set and functional testing on the free energies (kcal/mol) of formation of II_{Rh}, III_{Rh} and VII_{Rh-3} from I_{Rh} and $H-L_3$. All energies include a PCM correction for MeOH solvent

Entry	Functional/basis set	II_{Rh}	III_{Rh}	VII_{Rh-3}	ΔG^a
1	BP86/BS1[b]	−25.4	−46.5	−27.1	+19.4
2	BP86/BS2[c]	−8.6	−5.1	−1.3	+7.3
3	BP86-D3/BS2	−1.3	−0.4	−5.3	−4.0
4	B3PW91-D3/BS2	−1.5	−0.7	−4.2	−2.7
5	PBE-D3/BS2	−3.3	+1.3	−3.5	−0.2
6	PBE0-D3/BS2	−2.9	−1.9	−3.4	−0.5
7	BLYP-D3/BS2	−2.8	−1.9	−2.0	+0.8
8	B3LYP-D3/BS2	−3.2	−3.2	−1.5	+1.7
9	M06/BS2	−2.1	−0.6	+0.4	+2.5
10	M06L/BS2	−7.2	−10.6	+2.1	+12.7
11	ωB97XD	−6.7	−7.3	−5.2	+2.1
12	B97D/BS2	−4.8	−5.2	−7.3	−2.1

[a] ΔG: difference between the most stable precursor and VII_{Rh-3}
[b] BS1: Rh, S: SDD (polarization on S), C, H, N, O: 6-31G(d,p)
[c] BS2: Rh: cc-pVTZ-PP; S, C, H, N, O: 6-311++G(d,p)

be the only species observed in solution and so ruled out any possibility of cyclometallation. The situation improves somewhat when a larger basis set with diffuse functions on the ligand atoms was used (BS2, Table 1 Entry 2): II_{Rh} and III_{Rh} are now closer in energy, but are still significantly more stable than I_{Rh}; in addition, the formation of VII_{Rh-3} from II_{Rh} remains endergonic by 7.3 kcal/mol.

A treatment of dispersion effects proved essential to improve this situation. The use of Grimme's D3 parameter set (BP86-D3/BS2, Entry 3) placed I_{Rh}, II_{Rh} and III_{Rh} within 1.3 kcal/mol and showed cyclometallation to VII_{Rh-3} to be exergonic by 4.0 kcal/mol. The greater relative stabilization of VII_{Rh-3} reflects the proximity of the bulky N-^{i}Pr imine to the {Cp*Rh} fragment resulting in significant stabilizing dispersion interactions. This result also reiterates the importance of using the full experimental system in the calculations, as such stabilization will be greatly reduced with truncated model systems. Similar improved energetics were found with a range of other functionals, although BLYP-D3 and B3LYP-D3 along with M06, M06L and ωB97XD all underestimate the relative stability of VII_{Rh-3}.

The key energetics (using the BP86-D3/BS2 protocol) for the full reactions of H-L_3 at $[Cp*RhCl_2]_2$ (in MeOH) and $[Cp*IrCl_2]_2$ (in CH_2Cl_2) are summarized in Fig. 9. With $[Cp*RhCl_2]_2$ in MeOH, the reaction with H-L_3 proceeds with $\Delta G_{calc} = -4.0$ kcal/mol and $\Delta G^{\ddagger}_{calc} = 11.8$ kcal/mol, indicating reversible C–H activation proceeding under thermodynamic control. In contrast, the reaction of $[Cp*IrCl_2]_2$ with H-L_3 in CH_2Cl_2 is far more exergonic ($\Delta G_{calc} = -14.2$ kcal/mol) and also has a larger barrier ($\Delta G^{\ddagger}_{calc} = 18.4$ kcal/mol), consistent with irreversible C–H activation and kinetic control. Thus, the key behaviours seen experimentally are well reproduced. Note that the key conclusions depend primarily on the

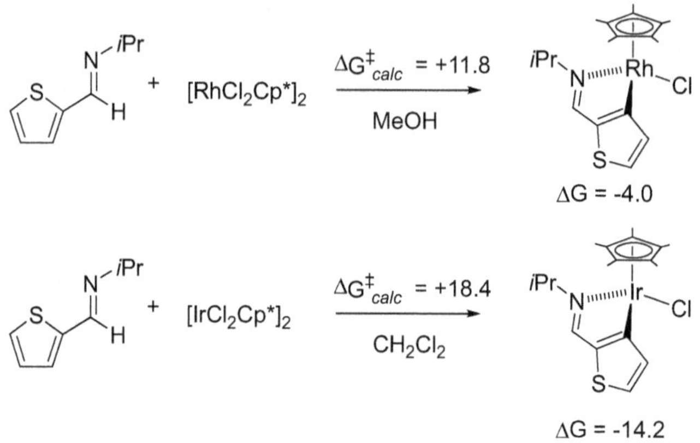
N
iPr
S
H
+
[RhCl2Cp*]2
ΔG‡calc = +11.8
MeOH
iPr
N
Rh
Cl
S
ΔG = -4.0
N
iPr
S
H
+
[IrCl2Cp*]2
ΔG‡calc = +18.4
CH2Cl2
iPr
N
Ir
Cl
S
ΔG = -14.2

Fig. 9 Computed key energetics (kcal/mol) for the cyclometallation of **H-L$_3$** at $[Cp^*MCl_2]_2$ (M = Rh and Ir; BP86-D3/BS2 level; see Table 1)

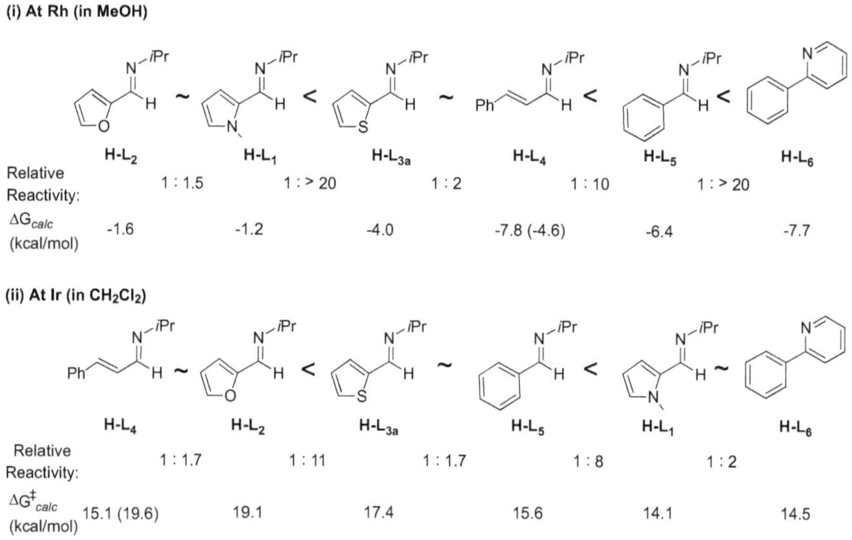
(i) At Rh (in MeOH)
H-L2 ~ H-L1 < H-L3a ~ H-L4 < H-L5 < H-L6
Relative Reactivity:
1 : 1.5
1 : > 20
1 : 2
1 : 10
1 : > 20
ΔGcalc (kcal/mol)
-1.6
-1.2
-4.0
-7.8 (-4.6)
-6.4
-7.7
(ii) At Ir (in CH2Cl2)
H-L4 ~ H-L2 < H-L3a ~ H-L5 < H-L1 ~ H-L6
Relative Reactivity:
1 : 1.7
1 : 11
1 : 1.7
1 : 8
1 : 2
ΔG‡calc (kcal/mol)
15.1 (19.6)
19.1
17.4
15.6
14.1
14.5

Fig. 10 Relative experimental and computed reactivities for substrates **H-L$_{1–6}$** at $[Cp^*MCl_2]_2$ (M = Rh and Ir). Computed data (kcal/mol) give the overall free energy changes, ΔG_{calc}, for M = Rh and calculated activation barriers, and $\Delta G^{\ddagger}_{\mathrm{calc}}$, for M = Ir

thermodynamic stabilities of the neutral chloride adducts **VII$_{\mathbf{M\text{-}3}}$** which are not particularly solvent dependent.

The relative reactivities computed for the six substrates at both **I$_{\mathbf{Rh}}$** (in terms of thermodynamic control, ΔG_{calc}) and **I$_{\mathbf{Ir}}$** (in terms of kinetic control, $\Delta G^{\ddagger}_{\mathrm{calc}}$) are shown in Fig. 10 alongside the orders derived from competition experiments. In general, good qualitative agreement is seen. Thus, for Rh, ΔG_{calc} generally becomes more exergonic from $\mathbf{H\text{-}L_2}$ to $\mathbf{H\text{-}L_6}$. The exceptions are (1) between $\mathbf{H\text{-}L_2}$ and $\mathbf{H\text{-}L_1}$ (although $\Delta\Delta G_{calc}$ here is only 0.4 kcal/mol, reflecting the similar reactivities of these substrates) and, more significantly, (2) for $\mathbf{H\text{-}L_4}$, where the value of the ΔG_{calc} would rank it as the most reactive substrate. A similar situation is seen for Ir, where the general decrease in $\Delta G^{\ddagger}_{calc}$ from $\mathbf{H\text{-}L_2}$ to $\mathbf{H\text{-}L_6}$ mirrors the experimental trend towards higher reactivity. However, once again a major exception arises for $\mathbf{H\text{-}L_4}$, which has an anomalously low barrier and, as a consequence, an overestimated reactivity.

Our interpretation of these discrepancies between the computed and experimental reactivity trends rests on the different size of the various substrates involved and how well our chemical model is able to capture solute–solvent dispersion interactions. In our current approach, the isolated substrates are taken from the solution and bound to a $\{Cp^*M(OAc)\}^+$ metal fragment. The D3 correction will capture the *intramolecular* dispersion interactions that arise upon adduct formation, and the degree of this stabilization will therefore reflect the size of the substrate. Imine $\mathbf{H\text{-}L_4}$ features a vinyl CH=CHPh substituent and therefore has a different shape to $\mathbf{H\text{-}L_{1-3}}$ and $\mathbf{H\text{-}L_5}$, which all have similar aromatic substituents. As a result, $\mathbf{H\text{-}L_4}$ apparently induces a greater dispersive stabilization upon binding to the metal fragment. This would not be an issue if correctly balanced by the dispersion stabilization between the substrates and the environment (i.e. the solvent), which in turn depends on the ability of the continuum solvation method to represent these interactions accurately. The discrepancies in the computed data indicate some shortcomings in the present approach. For $\mathbf{H\text{-}L_4}$, these effects are manifested in an exaggerated stability of all stationary points involving that substrate (i.e. $\Delta G^{\ddagger}_{calc}$ is underestimated at Ir and ΔG_{calc} is overestimated at Rh). To test this, we recomputed this system with a smaller CH=CH$_2$ substituent (values in parenthesis, Fig. 10), and indeed, this model gave both an increase in $\Delta G^{\ddagger}_{calc}$ at Ir (from 15.1 to 19.6 kcal/mol) and a reduction in ΔG_{calc} at Rh (from -7.8 to -4.6 kcal/mol). We also suspect a similar effect is in fact in play when comparing 2-phenylpyridine with the imines, all of which feature a bulky N-^{i}Pr group. This would cause the dispersion interactions to be underestimated with 2-phenylpyridine and would be consistent with the slightly lower barrier computed for $\mathbf{H\text{-}L_1}$ compared to $\mathbf{H\text{-}L_6}$ at Ir.

3.2.2   Balancing C–H Activation and Functionalization

Heteroatom-directed cyclometallation forms the basis of C–H functionalization strategies for heterocycle synthesis. An example is the Rh-catalysed oxidative coupling of 3-phenylpyrazoles (**1**) with 4-octyne (**2**) to give pyrazoloisoquinolines (**3**) [66]. The proposed catalytic cycle for this process is shown in Fig. 11 and was supported by DFT calculations (BP86) using both a simple model system [Model 1: 3-phenylpyrazole reacting with HC≡CH at CpRh(OAc)$_2$] and the full system used experimentally [Model 2: 3-phenyl-5-methylpyrazole reacting with 4-octyne at

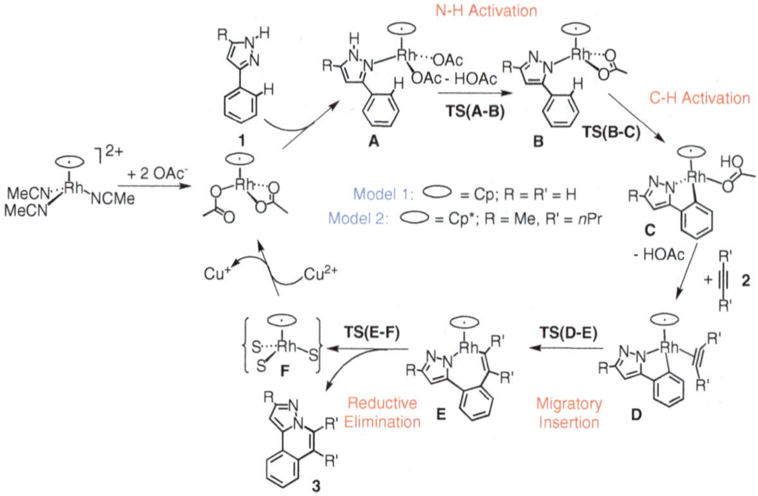

Fig. 11 General mechanism for Rh-catalysed formation of pyrazoloisoquinolines (**3**), via oxidative coupling of 3-phenylpyrazoles (**1**) and alkynes (**2**)

Cp*Rh(OAc)$_2$]. The key steps involved are the initial coordination and *N*-deprotonation of the 3-phenylpyrazole ligand (**A** → **B**), acetate-assisted C–H activation (**B** → **C**), HOAc/RC≡CR substitution to **D**, followed by migratory insertion to **E** and then reductive elimination to give the heterocyclic product. The Rh(I) species formed (loosely formulated as solvent adduct **F**) is reoxidized under the experimental conditions by Cu(OAc)$_2$ to regenerate the active Rh(III) catalyst.

This overall mechanism was not in dispute – and in the context of this chapter, it is worth noting that both the simple and the full models give the "same answer" in this respect, albeit with some changes in computed energies and geometries. However, the implications of experimental H/D exchange studies in CD$_3$OD proved more challenging to model computationally. These showed the C–H activation to be reversible, both in the absence of alkyne (40% incorporation at the *ortho* positions) and in the presence of alkyne (7% incorporation). Assuming facile H/D exchange of HOAc with the CD$_3$OD solvent in cyclometallated intermediate **C**, these observations indicate that there is a competition between the reprotonation of **C** on the one hand and the onward reaction via substitution with alkyne and migratory insertion on the other; i.e. **TS(B–C)** should be close in energy to **TS(D–E)**.

Our BP86 calculations (corrected for DCE solvent) are shown in Fig. 12a–c where in each case free energies are quoted relative to **A**, set to 0.0 kcal/mol. With this approach, Model 1 failed to capture the competition correctly, with **TS(D–E)** computed to lie 5.9 kcal/mol *below* **TS(B–C)** (see Fig. 12a). Adopting the full Model 2 was also initially unsuccessful as it now placed **TS(D–E)** 7.7 kcal/mol *above* **TS(B–C)** (Fig. 12b). As in the above examples, the situation only improved upon incorporating a treatment of dispersion effects via Grimme's D3 parameter

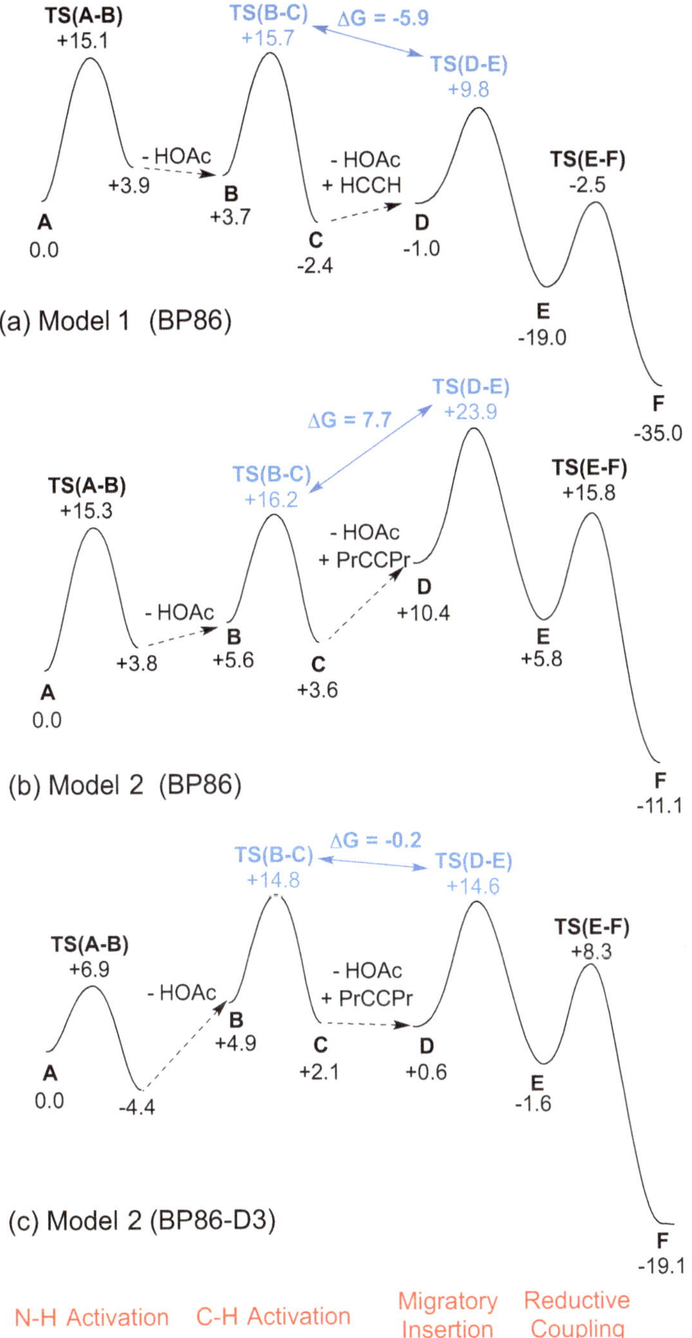

Fig. 12 Simplified computed free energy reaction profiles (kcal/mol) for the coupling of alkynes with 3-phenylpyrazoles in dichloroethane: (**a**) Model 1: $HC{\equiv}CH$ with **1b** at $CpRh(OAc)_2$, BP86 functional. (**b**) Model 2: $^nPrC{\equiv}C^nPr$ with **1a** at $Cp*Rh(OAc)_2$, BP86 functional (**c**) as for Model 2 but with BP86-D3

set: this stabilizes **TS(D–E)** ($\Delta G = +14.6$ kcal/mol), bringing it to within 0.2 kcal/ mol of **TS(B–C)** ($G = +14.8$ kcal/mol, Fig. 12c). Interestingly, this better agreement reflects changes in the energetics of the HOAc/^{n}PrC$\equiv$C^{n}Pr substitution step (**C** $\rightarrow$ **D**) rather than the migratory insertion step itself. The ligand substitution is highly sensitive to dispersion effects due to the different bulk of the two ligands involved: with BP86 this step is endergonic (by $+6.8$ kcal/mol), whereas with BP86-D3 it becomes exergonic (by -1.5 kcal/mol). The steric bulk around the Rh metal must also be correctly represented, i.e. Cp* must be employed and not just Cp; for this reason the truncated Model 1 fails even when dispersion effects are included. In contrast, the migratory insertion step (**D** $\rightarrow$ **E**) is not significantly affected by the dispersion correction, with similar barriers of 13.5 and 14.0 kcal/ mol computed with BP86 and BP86-D3, respectively. Thus, it is the stabilization of intermediate **D** when computed with dispersion that dictates the greater accessibility of **TS(D–E)** and hence allows the competition with **TS(B–C)** to be captured correctly. Alternative functionals that include a treatment of dispersion effects (e.g. M06 or B97D) also place **TS(D–E)** and **TS(B–C)** to within 1 kcal/mol of each other, and this behaviour was also computed when a D3 correction was added to a range of GGA and hybrid functionals.

3.3 Quantifying Barriers to Ph–I Bond Activation at Pd(0) and Ru(II)–Phosphine Complexes

The activation of aryl halides at low-valent Pd centres is a key step in the array of cross-coupling reactions that have been developed over the last 30–40 years. This process has therefore attracted considerable attention from computational chemists, and a significant body of work has arisen on Ar–X activation at Pd(PR$_3$)$_2$ species as reviewed recently [67]. As with the cyclopentadienyl family, the way in which phosphine ligands have been modelled in calculations has developed in response to the computing facilities and methodologies available, as well as the type of chemical problems that are being explored. Thus, in the 1980s and much of the 1990s, the essentially universal model for any phosphine was PH$_3$. The implementation of QM/MM methods allowed phosphine bulk to be considered (see also the chapter by Maseras and co-authors in this volume) [68, 69], while increased computational power saw the treatment of ever-larger phosphines with full DFT methods becoming increasingly routine. In comparing full QM (i.e. DFT) and QM/MM methods, it is important to note that the latter will capture both steric and dispersion effects but, depending on the approach, may or may not model substituent electronic effects. In contrast, a full DFT calculation will certainly capture the electronic effects of the substituent, while, as we have seen, the treatment of steric effects will be undermined by the inherent difficulties that DFT exhibits in handling long-range weak interactions. The treatment of phosphine dissociation energies therefore played a central role when the new methods including a treatment of dispersion began to be applied in organometallic chemistry [70–74]. A seminal paper in the context of Ar–X activation at $Pd(PR_3)_2$ species is that by Harvey, Fey and co-workers who demonstrated the importance of dispersion effects in correctly modelling the reactivity of the $Pd(P^tBu_3)_2$ moiety and in particular the role of mono-ligated 12-electron $Pd(P^tBu_3)$ [75]. While DFT calculations on transition metal systems containing several bulky phosphine ligands are now routinely accessible to computation, the study of such systems brings other issues, including an assessment of the conformational flexibility of the system. These various issues will be highlighted in the following two case studies.

3.3.1 Pd-Catalysed Azidocarbonylation of Aryl Iodides

The azidocarbonylation of aryl halides is a valuable transformation as the $ArC(O)$ N_3 products formed provide access to a range of useful isocyanate, urea and iminophosphorane precursors (Fig. 13). However, this very onward reactivity has severely limited the development of well-defined catalytic azidocarbonylation schemes as it places constraints on the temperature of reaction and emphasizes the need to avoid free phosphines in the reaction mixture. For Pd catalysis, this effectively limits the reaction to aryl iodides, the most readily activated of the aryl halides, while catalysts bearing phosphines tend to suffer from the onward Staudinger reaction which compromises the supporting phosphine ligand. Nevertheless, a successful catalytic system has recently been developed by Grushin and co-workers based on $(Xantphos)PdCl_2$ in the presence of a polymethylhydrosiloxane reductant and using biphasic conditions [76]. This system is capable of the room temperature gram-scale production of a range of $ArC(O)N_3$ products, with catalyst loadings as low as 0.2%. A combined experimental and computational study provided further insight into this process and defined a mechanism featuring rate-limiting Ph–I activation at a $(Xantphos)Pd(CO)_2$ active species. As in Sect. 3.2, a BP86-D3(BS2) protocol was employed, and the full system used experimentally was modelled in the calculations, with a correction for solvation in water.

Starting from tetrahedral $(Xantphos)Pd(CO)_2$, the calculations suggest that the reaction proceeds via initial loss of one CO ligand (see Fig. 14). The key intermediate is therefore $(Xantphos)Pd(CO)$, an insight that has ramifications for related carbonylation reactions that inevitably take place under a CO atmosphere (the reactive species may not necessarily be a $Pd(PR_3)_2$ complex as is often assumed).

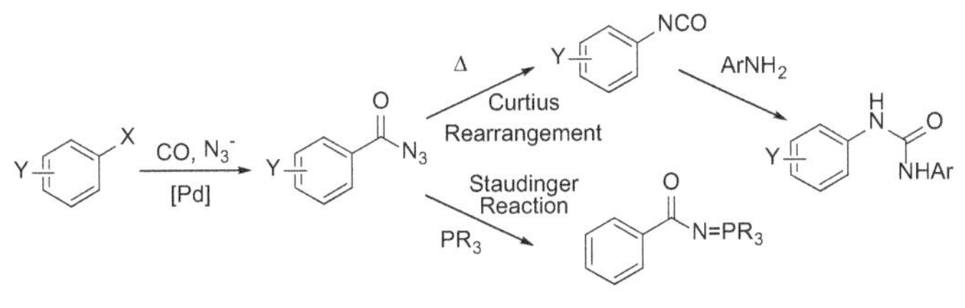

Fig. 13 Catalytic azidocarbonylation and onward reactivity

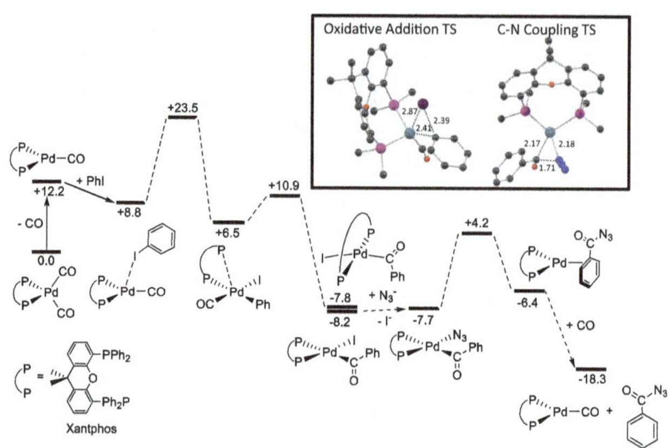

Fig. 14 Computed free energy reaction profile (kcal/mol) for catalytic azidocarbonylation at (Xantphos)Pd(CO)$_2$. *Inset* shows computed geometries of the oxidative addition and C–N coupling transition states, with key distances in Å

Once formed, (Xantphos)Pd(CO) reacts with Ph–I through a conventional three-centred oxidative addition transition state which the calculations locate at +23.5 kcal/mol. This forms a square-planar Pd(II) intermediate at +6.5 kcal/mol in which CO is *cis* to Ph and one arm of the Xantphos ligand has decoordinated. The reaction then proceeds via migratory insertion to give *trans*-(Xantphos)Pd({C(O)Ph}I, followed by *trans–cis* isomerization and anion exchange to give *cis*-(Xantphos)Pd({C(O)Ph}N$_3$ at −7.7 kcal/mol. C–N reductive coupling then readily occurs from this species with a barrier of only 11.9 kcal/mol. This sequence of events accounts for the features seen experimentally, in particular the fact that no Pd–azido intermediates could be observed: such species must be being formed in situ, but then rapidly proceed to the ArC(O)N$_3$ products via facile C–N bond coupling. An important observation from experiment was that the addition of NaN$_3$ to (Xantphos)Pd(Ph)I in a mixed organic/aqueous solvents sets up an equilibrium with the corresponding (Xantphos)Pd(Ph)N$_3$ and free iodide. ΔG for this equilibrium must therefore be close to zero, and this result was matched in the calculations once an extended basis set was used. Thus, the situation is similar to that described above when modelling Cl$^-$/OAc$^-$ exchange at Cp*RhCl(OAc) (see Sect. 3.2.1). Figure 14 indicates that the overall barrier for Ph–I activation at (Xantphos)Pd(CO)$_2$ is 23.5 kcal/mol and provides reasonable agreement with the value of 21.5 ± 0.5 kcal/mol determined experimentally for the reaction of 4-FC$_6$H$_4$I.

It is instructive to consider the steps that were necessary to achieve this good agreement between the experimental and computed activation energies for Ph–I activation. Figure 15 shows the energy changes associated with oxidative addition at (Xantphos)Pd(CO)$_2$ at the different levels of modelling employed. Entry 1 provides gas-phase free energies at the BP86/BS1 level and indicates the initial loss of CO costs 10.4 kcal/mol when computed with this approach. Adduct formation

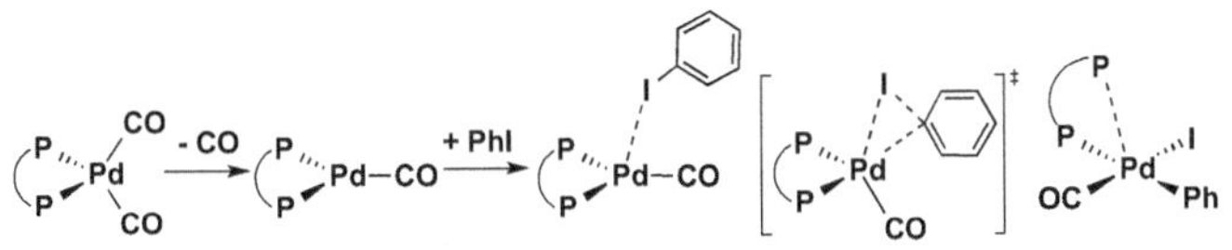

1. ΔG_{BS1}:	0.0	+10.4	+20.0	+41.1	+23.0
2. $\Delta G_{BS1}(H_2O)$:	0.0	+9.6	+20.2	+41.9	+20.1
3. $\Delta G_{BS1}(H_2O)$-D3:	0.0	+15.7	+11.4	+24.2	+8.8
4. $\Delta G_{BS2}(H_2O)$-D3:	0.0	+12.2	+8.8	+23.5	+6.5

Fig. 15 Evolution of the energy profile (BP86 with different protocols, kcal/mol) for the oxidative addition of PhI at (Xantphos)Pd(CO)$_2$. BS1: Pd, P: SDD (polarization on P), C,H,N,O: 6-31G(*d,p*); BS2: Pd: cc-pVTZ-PP; P, C, H, N, O: 6-311++G(*d,p*)

between (Xantphos)Pd(CO) and PhI entails a further destabilization of 9.6 kcal/mol. This implies only a weak interaction between these two components which fails to counter the significant entropic penalty caused by adduct formation. The barrier for the Ph–I cleavage step is then a further 21.1 kcal/mol meaning that the oxidative addition at (Xantphos)Pd(CO)$_2$ has a barrier of 41.1 kcal/mol and this step is endergonic by 23.0 kcal/mol. This outcome is clearly inconsistent with the room temperature azidocarbonylation seen experimentally. Inclusion of a correction for H$_2$O solvent had little effect on these values (Entry 2 in Fig. 15); however, the inclusion of the D3 dispersion correction proved dramatic. The general expectation is that dispersion will disfavour ligand dissociation and favour ligand association. Accordingly, CO dissociation from (Xantphos)Pd(CO)$_2$ rises to 15.7 kcal/mol (Entry 3, Fig. 15). However, this is now more than compensated by the stabilization of the (Xantphos)Pd(CO)·PhI adduct ($\Delta G = +11.4$ kcal/mol). Ph–I activation involves the approach of PhI towards the (Xantphos)Pd(CO) moiety, and so this is also favoured by the D3 correction meaning that the overall barrier is now reduced to only 24.2 kcal/mol. The final correction (Entry 4 in Fig. 15) is for the larger basis set, BS2, which is necessary to model the later anion exchange correctly: the effect here is rather small and reduces the barrier to the value of 23.5 kcal/mol reported above.

Conformation searching also played an important role in producing the above results. In this project, we were fortunate to have a number of crystal structures upon which the initial optimizations could be based. The Xantphos ligand in these and many related structures tends to adopt a symmetrical, near-C_s arrangement of the two {PPh$_2$} moieties, and this is also evident in the first version of the Ph–I activation transition state that we located in this study (conformer **9** in Fig. 16), the structure of which was derived from the experimental structure of (Xantphos)Pd(CO)$_2$. This had a relative free energy of 40.8 kcal/mol at the BP86/BS1 level but only dropped to 27.6 kcal/mol upon application of all corrections (solvent, dispersion and extended basis set, Entry 4). We therefore applied our conformational searching tool, which is based on high temperature constrained molecular dynamics to explore conformational space [77]. This revealed a further nine conformers, one

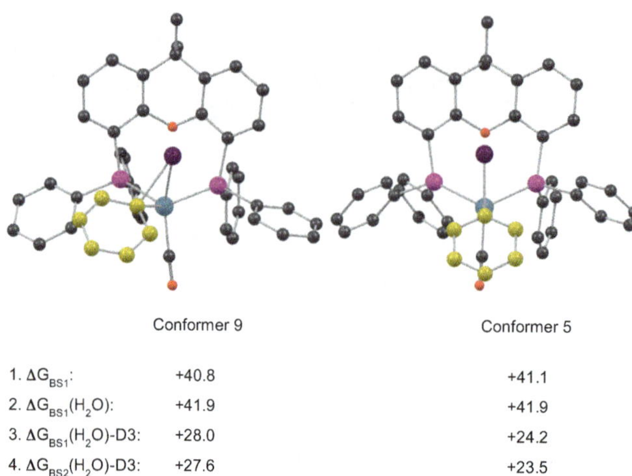

<table>
<tr><td></td><td>Conformer 9</td><td>Conformer 5</td></tr>
<tr><td>1. ΔG_{BS1}:</td><td>+40.8</td><td>+41.1</td></tr>
<tr><td>2. $\Delta G_{BS1}(H_2O)$:</td><td>+41.9</td><td>+41.9</td></tr>
<tr><td>3. $\Delta G_{BS1}(H_2O)$-D3:</td><td>+28.0</td><td>+24.2</td></tr>
<tr><td>4. $\Delta G_{BS2}(H_2O)$-D3:</td><td>+27.6</td><td>+23.5</td></tr>
</table>

Fig. 16 Evolution of the free energy (BP86 with different protocols, kcal/mol) of conformers **9** and **5** of the Ph–I oxidative addition transition state as a function of adopted methodology. Conformer **9** shows the symmetrical Xantphos ligand that is typical of crystallographic studies; conformer **5** is derived from an MD/MM search of conformational space. Atoms in *gold* highlight the phenyl ring of the PhI substrate, and H atoms have been omitted for clarity. See Fig. 15 for definition of BS1 and BS2

of which turned out to be relevant (conformer **5** in Fig. 16). At the BP86/BS1level, this new conformer was within 0.3 kcal/mol of the symmetric form, and these became isoenergetic when recomputed with the H_2O solvent correction. The D3 correction then discriminated between these two forms and had a greater effect on the nonsymmetrical conformer, placing it 3.8 kcal/mol below the symmetrical form (Entry 3 in Fig. 16). This difference then increased slightly to 4.1 kcal/mol at the final BP86-D3 (BS2, H_2O) level, giving the final free energy of +23.5 kcal/mol (Entry 4, Fig. 16) in reasonable agreement with the experimental value of 21.5 ± 0.5 kcal/mol. It is worth emphasizing that this good agreement was only achieved after a thorough assessment of the conformation space available to the system.

3.3.2 Ru-Mediated Ph–I Activation

Palladium dominates the field of transition metal-catalysed cross-coupling reactions. However, this metal is among the most expensive of the Pt group metals; moreover, in many cases intricate (hence costly) phosphine co-ligands may be required to ensure efficient and selective reactivity. The development of alternative catalysts therefore remains a priority. In comparison to Pd, Ar–X bond activation at Ru is poorly developed with known examples tending to require high temperatures or electron-rich alkylphosphines, factors which together bring no advantage over the established Pd chemistry. Recently, however, Grushin and co-workers have

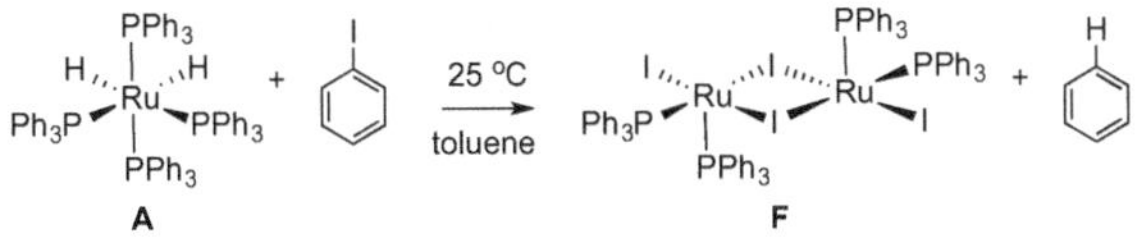

Fig. 17 Overall reaction of *cis*-[Ru(H)$_2$(PPh$_3$)$_4$], **A**, with PhI to give dimer [Ru(I)$_2$(PPh$_3$)$_2$]$_2$, **F**, and benzene

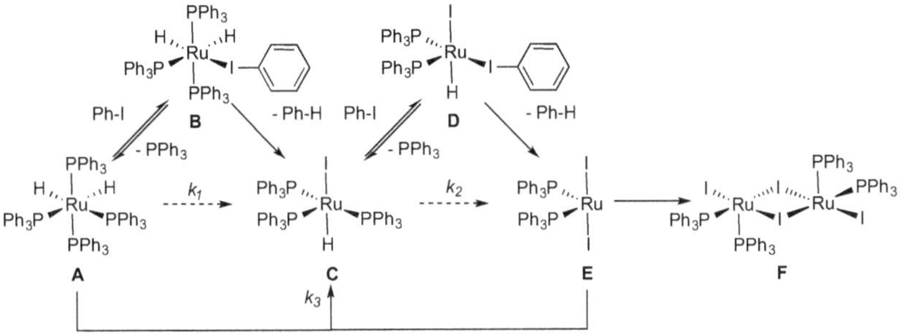

Fig. 18 Reaction sequence for the autocatalytic formation of C$_6$H$_6$ via Ph–I activation at *cis*-Ru (H)$_2$(PPh$_3$)$_4$, **A**, and Ru(H)(I)(PPh$_3$)$_3$, **C**

demonstrated the room temperature activation of Ph–I at a simple *cis*-Ru (PPh$_3$)$_4$(H)$_2$ species [78]. This system leads to the formation of benzene and ultimately an iodo-bridged dimer, [Ru(PPh$_3$)$_2$(I)$_2$]$_2$, **F** (Fig. 17). In addition, this system exhibits a remarkable autocatalytic behaviour that results in an *apparent* zeroth-order behaviour of the *cis*-Ru(PPh$_3$)$_4$(H)$_2$ reactant. This provided a distinct challenge for computational modelling, the results of which are described below.

Experimental mechanistic studies defined the reaction sequence shown in Fig. 18. In the presence of excess PhI, an equilibrium between Ru(PPh$_3$)$_4$(H)$_2$ (**A**) and the PPh$_3$-substituted species Ru(PPh$_3$)$_3$(H)$_2$(PhI) (**B**) is set up, where the Ph–I is bound through iodine. An equilibrium constant of $1.7 \pm 0.3 \times 10^{-3}$ was determined for this exchange reaction. Ph–I activation then proceeds to give Ru(PPh$_3$)$_3$(H) (I) (**C**) and benzene with an overall rate constant $k_1 = 1.24 \times 10^{-5}$ min^{-1}. Importantly, the Ru(PPh$_3$)$_3$(H)(I) formed in this step was found to activate Ph–I via **D**, generating a further equivalent of benzene and Ru(PPh$_3$)$_2$(I)$_2$ (**E**). The rate constant for this second C–I activation is $k_2 = 6.8 \times 10^{-6}$ min^{-1}, close to the value of k_1. Under the reaction conditions, the 14e species **E** forms during the second C–I activation and undergoes a rapid comproportionation reaction with the still-present **A** to regenerate **C**. The rate of this comproportionation, k_3, could not be determined but is thought to be considerably faster than the Ph–I activation reactions. These conditions, i.e. $k_3 \gg k_1 \approx k_2$, lead to an autocatalytic behaviour and the observed apparent zeroth-order kinetics in [Ru]. Once all of **A** is consumed, then **E** will dimerize to give **F**.

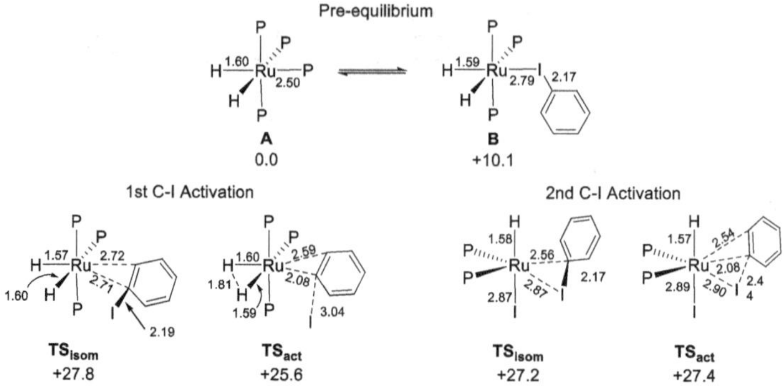

Pre-equilibrium

A 0.0 ⇌ **B** +10.1

1st C–I Activation

TS$_{isom}$ +27.8   **TS$_{act}$** +25.6

2nd C–I Activation

TS$_{isom}$ +27.2   **TS$_{act}$** +27.4

Fig. 19 Key stationary points computed for Ph–I activation at *cis*-[Ru(H)$_2$(P)$_4$], **A** (where P = PPh$_3$). Selected distances are given in Å, and free energies (kcal/mol) are computed with the BP86-D3(benzene) protocol and quoted relative to **A** + PhI set to 0.0 kcal/mol

The availability of quantitative experimental data provided an ideal opportunity to benchmark various DFT protocols. It is also rare that data are available on more than one step within a reaction sequence. The measured equilibrium constant between **A** and **B** is equivalent to $\Delta G_{eq} = 3.8$ kcal/mol in favour of **A**, while the rate constants k_1 and k_2 equate to barriers of $\Delta G_1^{\ddagger} = 26.6$ and $\Delta G_2^{\ddagger} = 26.9$ kcal /mol at room temperature for the first and second C–I activation processes, respectively. We focus here on the free energies of the relevant stationary points, namely, **A** and **B**, and the highest-lying transition states in each of the C–I activation process. Relevant geometries are shown in Fig. 19 and were optimized with the BP86/BS1 protocol used in the previous studies with free energies corrected for benzene solvent. After the pre-equilibrium involving PPh$_3$/PhI substitution *trans* to a hydride in **A**, isomerization of the PhI ligand to a π-bound form (via **TS$_{isom}$**) is required prior to the Ph–I activation (via **TS$_{act}$**). For the first C–I activation, **TS$_{isom}$** is slightly higher than **TS$_{act}$**. A similar substitution–isomerization–activation sequence is seen in the second C–I activation although in this case **TS$_{act}$** is higher than **TS$_{isom}$**. The differences here are small, and, indeed, which transition state is more accessible turns out to be functional dependent (see Fig. 20). The BP86-D3(benzene) protocol gives computed values of $\Delta G_1^{\ddagger} = 27.8$ kcal/mol via **TS$_{isom}$** and $\Delta G_2^{\ddagger} = 27.4$ kcal/mol via **TS$_{act}$** in excellent agreement with the values derived from experiment. In contrast, the calculated value for ΔG_{eq} between **A** and **B** is +10.1 kcal/mol, which is 6.3 kcal/mol larger than the experimental value.

The functional dependency of these results was then assessed, via a single point energy calculation at the BP86 geometry with the relevant D3 dispersion and benzene solvent correction included. The results are displayed in Fig. 20. The first five pairs of data contrast the behaviour of a range of pure and hybrid functionals, first without and then including the D3 correction. Once again, the

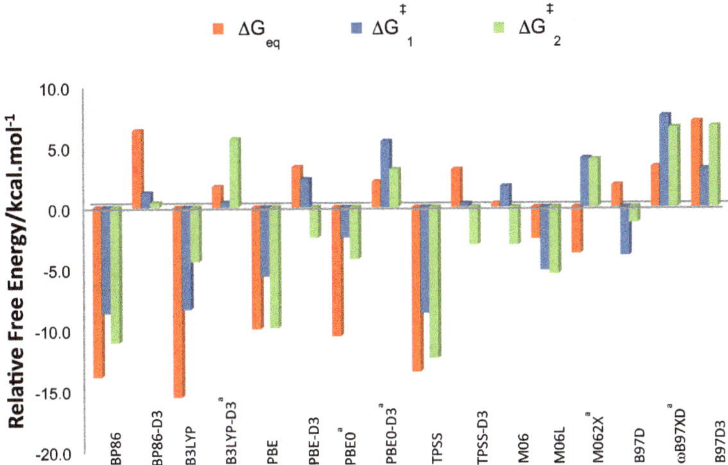

Fig. 20 Computed free energies (kcal/mol) for ΔG_{eq}, $\Delta G_1^{‡}$ and $\Delta G_2^{‡}$, displayed relative to the experimental data set to 0.0 kcal/mol. [a]TS_{act} lies above TS_{isom} (see Fig. 19 and associated text for details)

crucial role of dispersion is evident, as the uncorrected pure and hybrid functionals do not even give the correct sign for K_{eq} (i.e. **B** is predicted to be significantly more stable than **A**). This is improved by the equivalent D3-corrected results although the computed range (between +5.5 kcal/mol for B3LYP-D3 and +10.1 kcal/mol for BP86-D3) is always somewhat larger than the experimental value of $+3.8 \pm 0.1$ kcal/mol. Of the Minnesota functionals, M06 performed best in this respect, with a calculated K_{eq} of +4.1 kcal/mol. The B97 family of dispersion-corrected functionals also displays a range of behaviour with B97D being closest to experiment.

Experimentally, the barriers for the sequential C–I bond activation steps were the same within experimental error, and this was an important factor in producing the apparent autocatalytic behaviour. Of the protocols that include a treatment of dispersion effects that were assessed here, BP86-D3, PBE0-D3, M06L, M062X and ωB97XD provided overall barriers for these two steps that are within 2 kcal/mol of each other. For other functionals, significant differences can be seen (e.g. $\Delta G_2^{‡} - \Delta G_1^{‡}$: B3LYP-D3, +5.5 kcal/mol; M06, −4.5 kcal/mol; B97D3, +3.8 kcal/mol). In fact it is the original BP86-D3 approach that gives the best quantitative agreement with both $\Delta G_1^{‡}$ and $\Delta G_2^{‡}$ within 1.2 kcal/mol of experimental. Of the other functionals that provide similar values for $\Delta G_1^{‡}$ and $\Delta G_2^{‡}$, these are either rather too low (M06L: ca. 21 kcal/mol) or too high (M062X and ωB97XD, both in excess of 30 kcal/mol). The good performance of the BP86-D3 approach in reproducing experimental barriers was also seen above for the activation of Ph–I at Pd (Xantphos)(CO)$_2$. However, this method does not model the value of ΔG_{eq} well, with the computed value of +10.1 kcal/mol being over 6 kcal/mol above that from experiment. Other functionals, notably M06, B3LYP-D3 and B97D, perform much better in this regard, but none of these gives satisfactory barriers. It seems that no one functional is able to capture the overall behaviour of this system.

4 Conclusions

In this chapter, we posed the question "Modelling and Rationalizing Organometallic Chemistry with Computation. Where Are We?". The answer to this question depends upon where you want to be, i.e. on the level of information that is required! For qualitative information, e.g. for establishing key aspects of fundamental processes, simple models can be sufficient. Simple models can also help to determine a reference reaction coordinate which can be maintained or modified when including the complexity associated with the real systems. For instance, in Sect. 2, we saw how calculations consistently showed that σ-bond metathesis will involve transfer of a hydrogen in preference to transfer of an alkyl group. However, if we are focusing on a more quantitative aspect, for example, selectivity (which H would transfer?), or the absolute barrier for a certain process (in order to compare to experimental conditions), the situation becomes far more challenging. This often involves the use of large model systems, which then requires an appropriate treatment of the whole range of chemical interactions, from strong to weak, as well as tackling the high conformational diversity generally present in large systems. The case studies presented in Sect. 3 illustrate that, with care (and preferably with an open dialogue with experiment), a good representation of the experimental behaviour can indeed be achieved. This gives us optimism that predictions can be done when appropriate validation and analysis are in hand. Our overall philosophy is outlined in Fig. 21. More generally, challenges still remain such as solvation (particularly of charged species), dynamic behaviour, excited states and charge transfer, such as proton and/or electron transfers. Some of these are presented in other chapters in this volume and taken together capture the promise that the discipline of computational chemistry holds for the future. As time goes by,

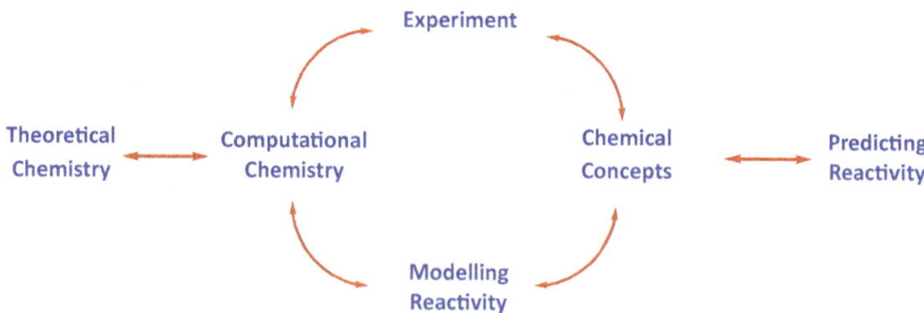

Fig. 21 Interrelationship between experimental, theoretical and computational chemistry

methods will improve and models will get ever closer to the real system, but the underlying philosophy described in this last figure will undoubtedly remain.

References

1. Woodward RB, Hoffmann R (1965) Stereochemistry of electrocyclic reactions. J Am Chem Soc 87:395–397
2. Woodward RB, Hoffmann R (1965) Selection rules for concerted cycloaddition reactions. J Am Chem Soc 87:2046–2048
3. Woodward RB, Hoffmann R (1965) Selection rules for sigmatropic sections. J Am Chem Soc 87:2511–2513
4. Woodward RB, Hoffmann R (1969) The conservation of orbital symmetry. Angew Chem Int Ed Engl 8:781–932
5. Fukui K (1982) The role of frontier orbitals in chemical reactions. Angew Chem Int Ed Engl 21:801–809
6. Hoffmann R (1982) Building bridges between inorganic and organic chemistry. Angew Chem Int Ed Engl 21:711–724
7. Cramer CJ, Truhlar DG (2009) Density functional theory for transition metals and transition metal chemistry. Phys Chem Chem Phys 11:10757–10816
8. Becke AD (2014) Perspective: fifty years of density-functional theory in chemical physics. J Chem Phys 140:18A301
9. Appelhans LN, Zuccaccia D, Kovacevic DA, Chianese AR, Miecznikowski JR, Macchioni A, Clot E, Eisenstein O, Crabtree RH (2005) An anion-dependent switch in selectivity results from a change of C-H activation mechanism in the reaction of an imidazolium salt with $IrH_5(PPh_3)_2$. J Am Chem Soc 127:16299–16311
10. Leduc A-M, Salameh A, Soulivong D, Chabanas M, Basset J-M, Copéret C, Solans-Monfort X, Clot E, Eisenstein O, Böhm VPW, Röper M (2008) β-H transfer from the metallacyclobutane: a key step in the deactivation and byproduct formation for the well-defined silica-supported rhenium alkylidene alkene metathesis catalyst. J Am Chem Soc 130:6288–6297
11. Erhardt S, Grushin VV, Kilpatrick AH, Macgregor SA, Marshall WJ, Roe DC (2008) Mechanisms of catalyst poisoning in palladium-catalyzed cyanation of haloarenes. Remarkably facile C-N bond activation in the $[(Ph_3P)_4Pd]/[Bu_4N]^+ CN^-$ system. J Am Chem Soc 130:4828–4845
12. Crabtree RH (2012) Resolving heterogeneity problems and impurity artifacts in operationally homogeneous transition metal catalysts. Chem Rev 112:1536–1554
13. Basch H, Musaev DG, Morokuma K, Fryzuk MD, Love JB, Seidel WW, Albinati A, Koetzle TF, Klooster WT, Mason SA, Eckert J (1999) Theoretical predictions and single-crystal neutron diffraction and inelastic neutron scattering studies on the reaction of dihydrogen with the dinuclear dinitrogen complex of zirconium $[P_2N_2]Zr(\mu-\eta^2-N_2)Zr[P_2N_2]$, $P_2N_2 = PhP$ $(CH_2SiMe_2NSiMe_2CH_2)_2PPh$. J Am Chem Soc 121:523–528
14. Lauher JW, Hoffmann R (1976) Structure and chemistry of bis(cyclopentadienyl)-MLn complexes. J Am Chem Soc 98:1729–1742
15. Thomas JR, Quelch GE, Seidl ET, Schaefer HF III (1992) The titane molecule (TiH_4): equilibrium geometry, infrared and Raman spectra of the first spectroscopically characterized transition metal tetrahydride. J Chem Phys 96:6857–6861
16. Maron L, Eisenstein O, Alary F, Poteau R (2002) Modeling C5H5 with atoms or effective group potential in lanthanide complexes: isolobality not the determining factor. J Phys Chem A 106:1797–1801
17. Steigerwald ML, Goddard WA III (1984) 2s + 2s Reactions at transition metals. 1. The reactions of D2 with Cl2TiH+, Cl2TiH, and Cl2ScH. J Am Chem Soc 106:308–311

18. Maron L, Eisenstein O (2000) Do f Electrons play a role in the lanthanide–ligand bonds? A DFT study of $Ln(NR_2)_3$; R = H, SiH_3. J Phys Chem A 104:7140–7143
19. Castro L, Kefalidis CE, McKay D, Essafi S, Perrin L, Maron L (2014) Theoretical treatment of one electron redox transformation of small molecule using f-element complexes. Dalton Trans 43:12124–12134
20. Maron L, Eisenstein O (2001) DFT Study of H–H Activation by Cp_2LnH d^0 Complexes. J Am Chem Soc 123:1036–1039
21. Maron L, Perrin L, Eisenstein O (2002) DFT study of CH_4 activation by d^0 Cl_2LnZ (Z = H, CH_3) complexes. J Chem Soc Dalton Trans 534–539
22. Perrin L, Maron L, Eisenstein O (2002) A DFT study of SiH_4 activation by Cp_2LnH. Inorg Chem 41:4355–4362
23. Perrin L, Maron L, Eisenstein O (2004) Lanthanide complexes: electronic structure and H-H, C-H and Si-H bond activation from a DFT perspective. In: Goldberg KI, Goldman AS (eds) Activation and functionalization of C-H bonds. ACS Symposium Series 885, pp 116–133
24. Eisenstein O, Maron L (2002) DFT studies of some structures and reactions of lanthanides complexes. J Organomet Chem 647:190–197
25. Barthelat JC, Chaudret B, Daudey JP, De Loth P, Poilblanc R (1991) Theoretical calculations on Nb and Ta trihydride complexes. Relations with the problem of quantum mechanical exchange coupling. J Am Chem Soc 113:9896–9898
26. Camanyes S, Maseras F, Moreno M, Lledós A, Lluch JM, Bertrán J (1996) Theoretical study of the hydrogen exchange coupling in the metallocene trihydride complexes $[(C_5H_5)_2MH_3]^{n+}$ (M = Mo, W, **n** =1; M = Nb, Ta, n = 0). J Am Chem Soc 118:4617–4621
27. Hyla-Kryspin I, Silverio SJ, Niu S, Gleiter R (1997) An ab initio investigation of σ-bond metathesis and insertion reactions of acetylene with Cl_2ZrH^+ and $Cl_2ZrCH_3^+$. J Mol Cat A Chem 115:183–192
28. Folga E, Ziegler T (1992) A theoretical study on the activation of hydrogen–hydrogen and hydrogen–alkyl bonds by electron-poor early transition metals. Can J Chem 70:333–342
29. Watson PL, Parshall GW (1985) Organolanthanides in catalysis. Acc Chem Res 18:51–56
30. Zachmanoglou CE, Docrat A, Bridgewater BM, Parkin G, Brandow CG, Bercaw JE, Jardine CN, Lyall M, Green JC, Keister JB (2002) The electronic influence of ring substituents and ansa bridges in zirconocene complexes as probed by infrared spectroscopic, electrochemical and computational studies. J Am Chem Soc 124:9525–9546
31. Ziegler T, Folga E, Berces A (1993) A Density functional study on the activation of hydrogen-hydrogen and hydrogen-carbon bonds by Cp_2Sc-H and Cp_2Sc-CH_3. J Am Chem Soc 115:636–646
32. Barros N, Eisenstein O, Maron L (2006) DFT studies of the methyl exchange reaction between Cp_2M–CH_3 or Cp^*_2M-CH_3 (Cp = C_5H_5, Cp^* = C_5Me_5, M = Y, Sc, Ln) and CH_4. Does M ionic radius control the reaction? Dalton Trans 3052–3057
33. Ziegler T, Folga E (1994) A density functional study on σ-bond metathesis reactions of possible importance in dehydrogenative silane polymerization. J Organomet Chem 478:47–65
34. Perrin L, Eisenstein O, Maron L (2007) Chemoselectivity in σ bond activations by lanthanocene complexes from a DFT perspective: reactions of Cp_2LnR (R = CH_3, H, SiH_3) with SiH_4 and CH_3-SiH_3. New J Chem 31:549–555
35. Perrin L, Maron L, Eisenstein O, Tilley TD (2009) Bond activations of $PhSiH_3$ by Cp_2SmH: a mechanistic investigation by the DFT method. Organometallics 28:3767–3775
36. Werkema EL, Messines E, Perrin L, Maron L, Eisenstein O, Andersen RA (2005) Hydrogen for fluorine exchange in $CH_{4-x}F_x$ by monomeric $[1,2,4-(Me_3C)_3C_5H_2]_2CeH$: experimental and computational studies. J Am Chem Soc 127:7781–7795
37. Maron L, Perrin L, Eisenstein O (2003) CF_4 defluorination by Cp_2Ln-H: a DFT study. Dalton Trans 22:4313–4318
38. Werkema EL, Andersen RA, Yahia A, Maron L, Eisenstein O (2009) Hydrogen for X-group exchange in CH_3X (X = Cl, Br, I, OMe, and NMe_2) by monomeric $[1,2,4-(Me_3C)_3C_5H_2]_2CeH$:

experimental and computational support for a carbenoid mechanism. Organometallics 28: 3173–3185

39. Werkema EL, Yahia A, Maron L, Eisenstein O, Andersen RA (2010) Bridging silyl groups in σ-bond metathesis and [1,2]-shifts. Experimental and computational study of the reaction between cerium metallocenes and MeOSiMe$_3$. Organometallics 29:5103–5110

40. Werkema EL, Yahia A, Maron L, Eisenstein O, Andersen RA (2010) Splitting a C–O bond in dialkylethers with bis(1,2,4-tri-tert-butylcyclopentadienyl)cerium does not occur by a σ-bond pathway: a combined experimental and DFT computational study. New J Chem 34:2189–2196

41. Werkema EL, Castro L, Maron L, Eisenstein O, Andersen RA (2013) Cleaving bonds in CH$_3$OSO$_2$CF$_3$ with [1,2,4-(Me$_3$C)$_3$C$_5$H$_2$]$_2$CeH; an experimental and computational study. New J Chem 37:132–142

42. Werkema EL, Castro L, Maron L, Eisenstein O, Andersen RA (2012) Selectivity in the C–H activation reaction of CH$_3$OSO$_2$CH$_3$ with [1,2,4-(Me$_3$C)$_3$C$_5$H$_2$]$_2$CeH or [1,2,4-(Me$_3$C)$_3$C$_5$H$_2$][1,2-(Me$_3$C)$_2$-4-(Me$_2$CCH$_2$)C$_5$H$_2$]Ce: to choose or not to choose. Organometallics 31: 870–881

43. Maron L, Werkema EL, Perrin L, Eisenstein O, Andersen RA (2005) Hydrogen for fluorine exchange in C$_6$F$_6$ and C$_6$F$_5$H by monomeric [1,3,4-(Me$_3$C)$_3$C$_5$H$_2$]$_2$CeH: experimental and computational studies. J Am Chem Soc 127:279–292

44. Werkema EL, Andersen RA (2008) Fluorine for hydrogen exchange in the hydrofluorobenzene derivatives C$_6$H$_x$F$_{(6-x)}$, where $x = 2$, 3, 4 and 5 by monomeric [1,2,4-(Me$_3$C)$_3$C$_5$H$_2$]$_2$CeH: the solid state isomerization of [1,2,4-(Me$_3$C)$_3$C$_5$H$_2$]$_2$Ce(2,3,4,5-C$_6$HF$_4$) to [1,2,4-(Me$_3$C)$_3$C$_5$H$_2$]$_2$Ce(2,3,4,6-C$_6$HF$_4$). J Am Chem Soc 130:7153–7165

45. McKay D, Riddlestone IM, Macgregor SA, Mahon MF, Whittlesey MK (2015) A mechanistic study of Ru-NHC catalysed hydrodefluorination of fluoropyridines: the influence of the NHC on the regioselectivity of C-F activation and chemoselectivity of C-F vs C-H bond cleavage. ACS Catal 5:776–787

46. Werkema EL, Andersen RA, Maron L, Eisenstein O (2010) The reaction of bis(1,2,4-tri-t-butylcyclopentadienyl)ceriumbenzyl, Cp$'_2$CeCH$_2$Ph, with methylhalides: a metathesis reaction that does not proceed by a metathesis transition state. Dalton Trans 39:6648–6660

47. Scherer E, Cramer CJ (2003) Quantum chemical characterization of methane metathesis in L$_2$MCH$_3$ (L = H, Cl, Cp, Cp*; M = Sc, Y, Lu). Organometallics 22:1682–1689

48. Woodrum NL, Cramer CJ (2006) Density functional characterization of methane metathesis with Cp*$_2$MR (M = Sc, Y, Lu; R = Me, ^{t}BuCH$_2$). Structural and kinetic consequences of alkyl steric bulk. Organometallics 25:68–73

49. Barros N, Eisenstein O, Maron L, Tilley TD (2006) DFT investigation of the catalytic hydromethylation of α-olefins by metallocenes. 1. Differences between scandium and lutetium in propene hydromethylation. Organometallics 25:5699–5708

50. Barros N, Eisenstein O, Maron L, Tilley TD (2008) DFT investigation of the catalytic hydromethylation of olefins by scandocenes. 2. Influence of the ansa ligand on propene and isobutene hydromethylation. Organometallics 27:2252–2257

51. Barros N, Eisenstein O, Maron L (2010) Catalytic hydrosilylation of olefins with organo-lanthanides: a DFT study. Part I: hydrosilylation of propene by SiH$_4$. Dalton Trans 39: 10749–10756

52. Barros N, Eisenstein O, Maron L (2010) Catalytic hydrosilylation of olefins with organo-lanthanide complexes: a DFT study. Part II: influence of the substitution on olefin and silane. Dalton Trans 39:10757–10767

53. Harvey JN (2010) Ab initio transition state theory for polar reactions in solution. Faraday Discuss 145:487–505

54. Plata RE, Singleton DA (2015) A case study of the mechanism of alcohol-mediated Morita-Baylis-Hillman reactions. The importance of experimental observations. J Am Chem Soc 137: 3811–3826

55. Perrin L, Werkema EL, Eisenstein O, Andersen RA (2014) Two [1,2,4-(Me$_3$C)$_3$C$_5$H$_2$]$_2$CeH molecules are involved in the hydrogenation of pyridine to piperidine as shown by experiments and calculations. Inorg Chem 53:6361–6373

56. Deelman BJ, Stevels WM, Teuben JH, Lakin MT, Spek AL (1994) Insertion chemistry of yttrium complex Cp*$_2$Y(2-pyridyl) and molecular structure of an unexpected CO insertion product (Cp*$_2$Y)$_2$(μ-η^2:η^2-OC(NC$_5$H$_4$)$_2$). Organometallics 13:3881–3891

57. Evans WJ, Meadows JH, Hunter WE, Atwood JL (1984) Organolanthanide and organoyttrium hydride chemistry. 5. Improved synthesis of [(C$_5$H$_4$R)$_2$YH(THF)]$_2$ complexes and their reactivity with alkenes, alkynes, 1,2-propadiene, nitriles, and pyridine, including structural characterization of an alkylideneamido product. J Am Chem Soc 106:1291–1300

58. Werkema EL, Maron L, Eisenstein O, Andersen RA (2007) Reactions of monomeric [1,2,4-(Me3C)3C5H2]2CeH and CO with or without H2: an experimental and computational study. J Am Chem Soc 129:2529–2541

59. Grimme S, Ehrlich S, Goerigk L (2011) Effect of the damping function in dispersion corrected density functional theory. J Comput Chem 32:1456–1465

60. Grimme S, Steinmetz M (2013) Effects of London dispersion correction in density functional theory on the structures of organic molecules in the gas phase. Phys Chem Chem Phys 15:16031–16042

61. Hansen A, Bannwarth C, Grimme S, Petrović P, Werlé C, Djukic J-P (2014) The thermochemistry of london dispersion-driven transition metal reactions: getting the 'right answer for the right reason'. ChemistryOpen 3:177–189

62. Brandenburg JG, Bender G, Ren J, Hansen A, Grimme S, Eckert H, Daniliuc CG, Kehr G, Erker G (2014) Crystal packing induced carbon–carbon double–triple bond isomerization in a zirconocene complex. Organometallics 33:5358–5364

63. Moellmann J, Grimme S (2013) Influence of crystal packing on an organometallic ruthenium (IV) complex structure: the right distance for the right reason. Organometallics 32:3784–3787

64. Carr KJT, Davies DL, Macgregor SA, Singh K, Villa-Marcos B (2014) Metal control of selectivity in acetate-assisted C-H bond activation: an experimental and computational study of heterocyclic, vinylic and phenylic C(sp^2)-H bonds at Ir and Rh. Chem Sci 5:2340–2346

65. Ling L, Brennessel WW, Jones WD (2009) C–H activation of phenyl imines and 2-phenylpyridines with [Cp*MCl$_2$]$_2$ (M = Ir, Rh): regioselectivity, kinetics, and mechanism. Organometallics 28:3492–3500

66. Algarra AG, Cross WB, Davies DL, Khamker Q, Macgregor SA, McMullin CL, Singh K (2014) Combined experimental and computational investigations of rhodium- and ruthenium-catalyzed C-H functionalization of pyrazoles with alkynes. J Org Chem 79:1954–1970

67. Algarra AG, Macgregor SA, Panetier JA (2013) Mechanistic studies of C-X bond activation at transition-metal centers. In: Reedijik J, Poeppelmeier K (eds) Comprehensive inorganic chemistry II, vol 9. Elsevier, Amsterdam, pp 635–694

68. Svensson M, Humbel S, Froese RJD, Matsubara T, Sieber S, Morokuma K (1996) ONIOM: a multilayered integrated MO + MM method for geometry optimizations and single point energy predictions. A test for Diels-Alder reactions and Pt(P(^{t}Bu)$_3$)$_2$ + H$_2$ oxidative addition. J Phys Chem 100:19357–19363

69. Bo C, Maseras F (2008) QM/MM methods in inorganic chemistry. Dalton Trans 2911–2919

70. Ahlquist MSG, Norrby PO (2011) Dispersion and back-donation gives tetracoordinate [Pd (PPh$_3$)$_4$]. Angew Chem Int Ed 50:11794–11797

71. Minenkov Y, Occhipinti G, Jensen VR (2009) Metal-phosphine bond strengths of the transition metals: a challenge for DFT. J Phys Chem A 113:11833–11844

72. Ryde U, Mata RA, Grimme S (2011) Does DFT-D estimate accurate energies for the binding of ligands to metal complexes? Dalton Trans 40:11176–11183

73. Sieffert N, Bühl M (2009) Noncovalent interactions in a transition-metal triphenylphosphine complex: a density functional case study. Inorg Chem 48:4622–4624

74. Zhao Y, Truhlar DG (2007) Attractive noncovalent interactions in the mechanism of Grubbs second-generation Ru catalysts for olefin metathesis. Org Lett 9:1967–1970

75. McMullin CL, Jover J, Harvey JN, Fey N (2010) Accurate modelling of Pd(0) + PhX oxidative addition kinetics. Dalton Trans 39:10833–10836
76. Miloserdov FM, McMullin CL, Belmonte MM, Benet-Buchholz J, Bakhmutov VI, Macgregor SA, Grushin VV (2014) The challenge of palladium-catalyzed aromatic azidocarbonylation: from mechanistic and catalyst deactivation studies to a highly efficient process. Organometallics 33:736–752
77. Häller LJL, Page MJ, Erhardt S, Macgregor SA, Mahon MF, Naser MA, Velez A, Whittlesey MK (2010) Experimental and computational investigation of C-N bond activation in ruthenium N-heterocyclic carbene complexes. J Am Chem Soc 132:18408–18416
78. Miloserdov FM, McKay D, Muñoz BK, Samouei H, Macgregor SA, Grushin VV (2015) Exceedingly facile Ph-X activation (X = F, Cl, Br, I) with Ru(II): arresting kinetics, autocatalysis, and mechanisms. Angew Chem Int Ed 6:8466–8470

Struct Bond (2016) 167: 39–58
DOI: 10.1007/430_2014_148
© Springer-Verlag Berlin Heidelberg 2014
Published online: 22 May 2014

Reactivities and Electronic Properties of Boryl Ligands

Zhenyang Lin

Abstract Transition metal boryl complexes play an important role as reactive intermediates in many catalytic processes. The reactivities and electronic properties of boryl ligands, in particular their nucleophilic and electrophilic behaviors, are discussed and reviewed in this article.

Keywords Boryl complexes · Boryl ligands · Borylation · Structure and bonding

Contents

1 Introduction

Transition metal boryl (M-BY$_2$, Y=OR, NR$_2$, alkyl, etc.) complexes have attracted considerable interest because they play important roles as reactive intermediates in many catalytic processes for the synthesis of organoboron compounds which are important reagents in organic chemistry enabling many chemical transformations [1–19]. Over the past few decades, tremendous progress has been made in the aspects of their syntheses, reactivities, and structures, and a number of reviews have

Z. Lin (✉)
Department of Chemistry, The Hong Kong University of Science and Technology, Clear Water Bay, Kowloon, Hong Kong
e-mail: chzlin@ust.hk

been published on this subject [3–9, 12–16, 18]. In a recent Feature article, we have discussed selected aspects of metal-boryl chemistry, pertaining especially to late and post-transition metal boryl (M-BY_2) complexes in various catalytic processes, and highlighted the nucleophilic behavior of the boryl ligands [14]. In view of recent developments influencing our understanding of boryl ligands, the need to update what have been presented previously is apparent. Therefore, the present article can be viewed as an extension of our Feature article for late and post-transition metal boryl complexes in which their structure and bonding aspects are particularly discussed.

2 Structural *trans*-Influence of Boryl Ligands

In a boryl ligand, the pair of electrons for ligand-to-metal σ bonding occupies a formally sp^2-hybridized orbital at the boron center (**1**). Because of the electropositive nature of boron, boryl ligands usually exhibit a very strong *trans* influence. Although the three-coordinate boron center in a boryl ligand possesses an "empty" p orbital which is capable of having a π-interaction with the transition metal center bonded to it, it is generally accepted that the π-interaction is relatively weak, especially in M-$B(OR)_2$ systems wherein the B–O π interaction seems to be dominant [20–26].

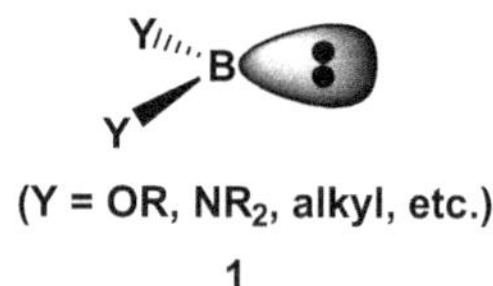

(Y = OR, NR$_2$, alkyl, etc.)

1

With the aid of DFT calculations, the *trans* influence of boryl ligands in a series of square-planar platinum(II) complexes of the form *trans*-[PtL(Cl)(PMe$_3$)$_2$] has previously been investigated [27]. Substituents R on the boryl ligands determine the percent s and p character in the formally sp^2-hybridized orbital. The more p-character the hybridized orbital has, the stronger the σ-donor strength and the greater *trans* influence is of the boryl ligand. Using the Pt-Cl distances as a measure, the following order of *trans* influence was obtained: –BMe$_2$ > –SiMe$_3$ > –BH$_2$ > –SnMe$_3$ ~ –BNHCH$_2$CH$_2$NH > –Bpin > –BOCH$_2$CH$_2$O > –BOCH=CHO ~ –Bcat ~ –BCl$_2$ ~ –BBr$_2$ ~ –SiH$_3$ > –CH$_2$CH$_3$ > –CH=CH$_2$ > –H ~ –Me > –C$_6$H$_5$ > –SiCl$_3$ > –SnCl$_3$ > –C≡CH (pin=pinacolato=OCMe$_2$CMe$_2$O; cat=catecholato=1,2-O$_2$C$_6$H$_4$).

Based on the Pt–Br bond distances of a series of structurally characterized platinum (II) monoboryl complexes of the general formula *trans*-[(Cy$_3$P)$_2$Pt(Br)(BX$_2$)], Braunschweig and co-workers obtained the following order of *trans* influence [28]: –B(Br)^{t}Bu > –B(Br)piperidenyl > –B(NMe$_2$)$_2$ > –B(Br)Mes > –B(Br)Fc > –B(Br)NMe$_2$ > –B(Br)(*o*-tolyl) > –BBr$_2$ > –Bcat′, where Fc = ferrocenyl and cat′ = cat-4-^{t}Bu.

Before continuing our discussion, we would like to refer readers to an early and a recent review article which provided comprehensive overview and discussions of the structural aspects of metal-boryl complexes based on the available X-ray crystallographic data [5, 13]. Herein, we mainly discuss how the very strong *trans* influence properties of boryl ligands determine the experimentally observed structural and geometric features of these transition metal boryl complexes. The discussion below is mainly based on the X-ray crystal structures of selected known transition boryl complexes.

The very strong *trans* influence of boryl ligands is manifested not only in the complexes mentioned above but also in the X-ray crystal structures of the complexes shown in Scheme 1. The monoboryl complexes **2**, **3**, and **4** are all 18-electron species [2, 29, 30]. The boryl ligand in each of the three complexes occupies a site *trans* to the weak *trans*-influencing chloride ligand.

The structures of the two square planar complexes **5** and **6** shown in Scheme 1 are also interesting [31, 32]. It is understandable that the boryl ligand in **5** is *trans* to the chloride ligand. The arrangement of the four ligands is optimal in that the strong and weak *trans* influence ligands are *trans* to each other and that the two sterically very bulky PPh_3 ligands are also *trans* to each other to avoid severe steric repulsion. This optimal arrangement has been found in a number of structurally characterized platinum(II) monoboryl complexes of the general formula *trans*-[$(Cy_3P)_2Pt(Br)$ (BX_2)] [28, 33–37].

A series of cationic, T-shaped, 14-electron boryl complexes of the type *trans*-[$(Cy_3P)_2Pt\{BXX'\}$]$^+$ (X=Cl, Br, $BNMe_2Br$; X'=ortho-tolyl, ^{t}Bu, NMe_2, piperidyl, ferrocenyl, Cl, Br; XX'=$(NMe_2)_2$, catecholato) by halide abstraction from *trans*-[$(Cy_3P)_2Pt(Br)\{B(X)X'\}$] have also been synthesized and structurally characterized [38–40]. In these complexes, the site *trans* to the boryl ligands is either vacant or has a very weak Pt–H–C(Cy) agostic bond due to the very strong *trans* influence of boryl ligands.

It is interesting to observe that the two boryl ligands are *cis* to each other in **6** despite the fact that the two triphenyl phosphine ligands are sterically very bulky. Clearly, the two very strong *trans* influence ligands avoid being in a *trans* arrangement. It should be noted here that **6** and its related square planar bis–boryl platinum complexes display unusual, acute B–Pt–B angles (in the range of 73–81°) [5, 31, 32, 41–46]. The origin of the acute angles in these square planar bis–boryl platinum complexes requires further investigation. Similar structural feature (acute B–M–B angles) has also been observed in other bis–boryl complexes which were considered to possess B–B interactions [47, 48].

In the bis-boryl complexes **7**, **8**, and **9** [49–52], the cis arrangements of the two boryl ligands prevent the very strong *trans* influence ligands from being *trans* to each other. In the tris–boryl complexes **10**, **11**, and **12** [53–55], the facial arrangements of the three boryl ligands do the same.

Scheme 2 shows several structurally characterized octahedral monoboryl complexes containing two very bulky PPh_3 ligands and one or more carbonyl ligands [50, 56, 57]. For those complexes with only one carbonyl ligand present (**13**, **14**, and **15**), the boryl ligand avoids being *trans* to the carbonyl group which is also a strong

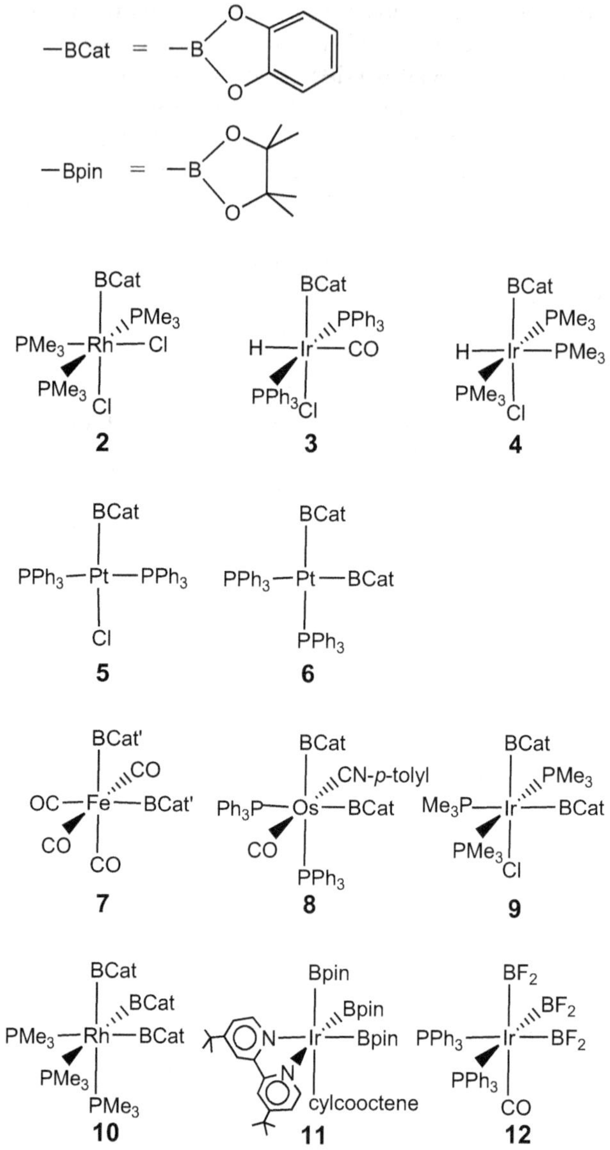

Scheme 1

trans influence ligand. Complexes **16** and **17**, which have two carbonyl ligands, are a pair of geometric isomers. In **16**, although the *trans* arrangement of the boryl and iodide ligands is optimal, that of the two carbonyl ligands is not. Thus, isomer **17** also exists. The same is true for the pair of geometric isomers **18** and **19**.

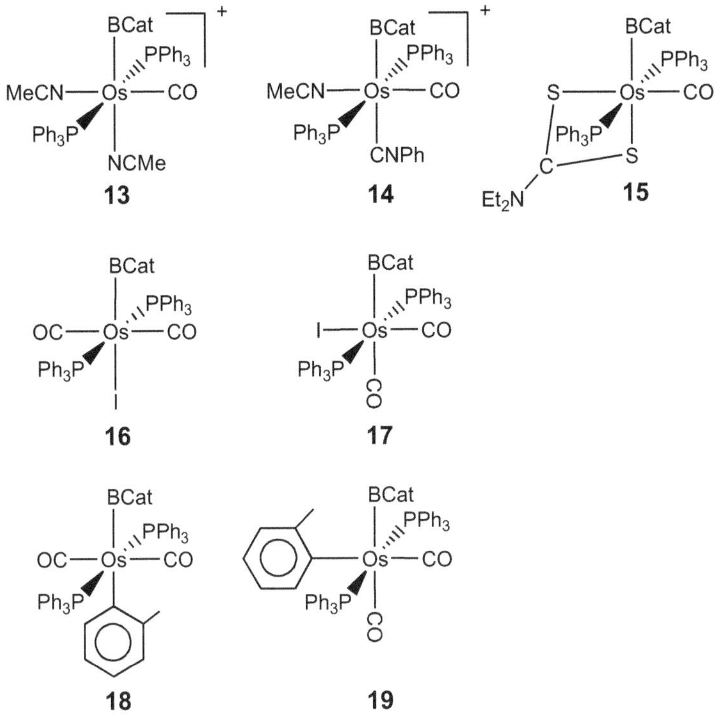

Scheme 2

The monoboryl complexes **20** and **21** shown in Scheme 3 adopt square pyramidal (SQP) geometries and are 16-electron species [52, 56, 58–60]. The optimal arrangement of the ligands is such that the strong *trans* influencing boryl ligand in each of the SQP complexes occupies the apical site because the site *trans* to it is vacant (Fig. 1). In complex **22** [61], the apical site is not occupied by a boryl ligand. This complex is a 17-electron species. The vacant site is not exactly "vacant." Instead, there is one electron occupying the molecular orbital that has its maximum amplitude in the direction of the vacant site. The singly occupied molecular orbital has significant M–L σ^*-antibonding character, preventing the very strong *trans* influence boryl ligand from occupying the apical site [62]. In the 18-electron trigonal bipyramidal (TBP) complex **23** [50], the boryl ligand avoids occupation of the equatorial sites in order to stabilize the orbitals that accommodate metal d electrons (Fig. 1) [62]. Thus, the boryl ligand occupies an axial site, as expected for the strongest σ-donor in a d^8-system [63], rather than an equatorial site which would be expected to be used by a strong π-acceptor ligand. This preference again highlights the fact that the boryl ligand's strong σ-donor properties seem to outweigh its π-acceptor properties.

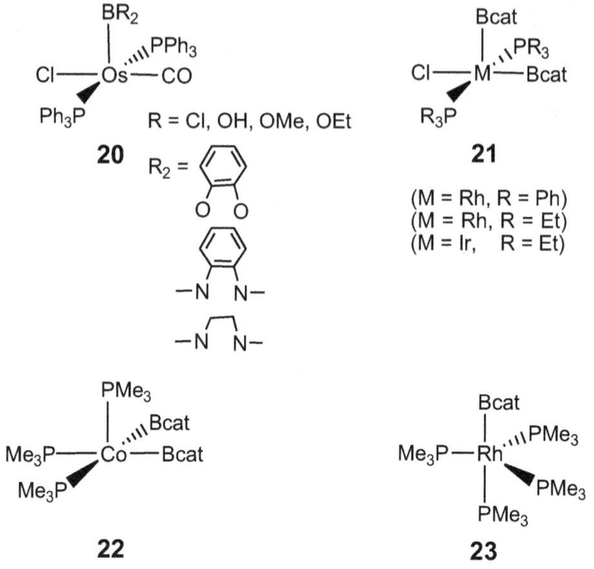

20

R = Cl, OH, OMe, OEt

21

(M = Rh, R = Ph)
(M = Rh, R = Et)
(M = Ir, R = Et)

22

23

Scheme 3

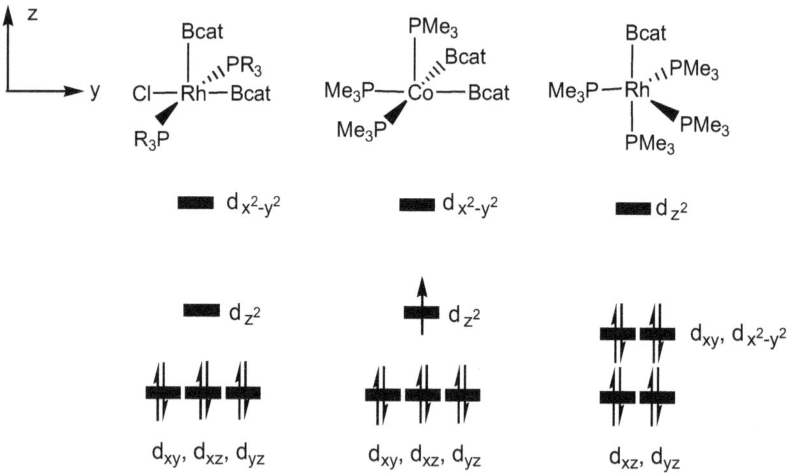

Fig. 1 Qualitative d orbital energy levels for three representative complexes shown in Scheme 3

3 Nucleophilic Properties of Boryl Ligands

In our previous Feature article [14], we discussed reaction mechanisms for various copper-catalyzed borylation processes and demonstrated that boryl ligands display nucleophilic behavior, in spite the presence of an "empty" p orbital on the boron center, as first suggested for certain copper promoted borylations [64, 65]. The reaction mechanisms discussed in the Feature article include the reduction of CO_2 with B_2pin_2 [66, 67], the diboration of aldehydes [68, 69] and the β-borylation of α,β-unsaturated carbonyl compounds [70–75] catalyzed by copper(I) complexes [Eqs. (1)–(3)]. In all of these reactions, a copper(I) boryl complex acts as the active species. Nucleophilic attack by the copper-bonded boryl ligand on the CO_2 carbon, an aldehyde carbonyl carbon and the β-carbon of an α,β-unsaturated carbonyl molecule initiates the catalytic reactions. In other words, the nucleophilicity, not the oxophilicity, of the boryl ligand determines the direction of the substrate insertion into the Cu–boryl bond.

$$CO_2 \ + \ B_2pin_2 \ \xrightarrow{\text{(NHC)Cu(Bpin)}} \ CO \ + \ (pinB)_2O \tag{1}$$

$$\tag{2}$$

ICy = 1,3-dicyclohexylimidazol-2-ylidene

$$\tag{3}$$

The nature of nucleophilicity of the boryl ligands in the active LCu(boryl) species is consistent with the fact that the HOMO calculated for the complex is mainly associated with the Cu–B σ-bonding interaction and the LUMO is related to the C=C π^*-antibonding orbital of the NHC ligand instead of the "empty" p orbital on the boron center (Fig. 2) [76]. The interaction of the active species with an organic substrate (aldehyde or α,β-unsaturated carbonyl) is closely related to the orbital interaction between the HOMO of the copper(I) boryl species and the LUMO of the substrate. It is worth noting that the HOMO is more localized on boron than on copper. Figure 3 gives frontier orbitals calculated for formaldehyde and acrolein [75]. In formaldehyde and acrolein, the preferred site being attacked by a boryl ligand is associated with the carbon which has the largest orbital character in its respective LUMO.

Another reaction discussed below further demonstrates the nucleophilic behavior of boryl ligands. Sadighi and co-workers investigated the insertion reaction of styrene with the *N*-heterocyclic carbene-ligated copper boryl complex **24**, shown in Scheme 4 [77]. The insertion product **25**, a β-borylalkyl complex, was isolated and structurally characterized by single-crystal X-ray diffraction. On heating, **25** undergoes a β-hydride elimination and then reinsertion sequence to afford the rearranged α-borylalkyl complex **26** (Scheme 4).

The DFT study showed that the free energy barrier (15.1 kcal/mol) calculated for the 2,1-insertion, which leads to the observed product **25**, is much lower than that

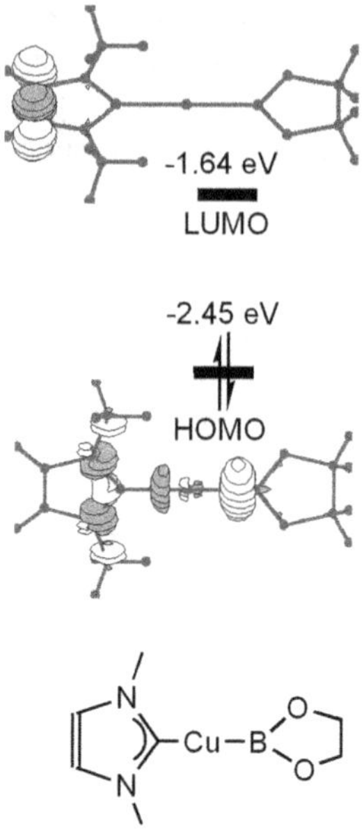

Fig. 2 Frontier molecular orbitals calculated for the model complex (NHC)Cu (boryl)

(24.7 kcal/mol) for the 1,2-insertion. Again, the 2,1-insertion product **25** is a result of nucleophilic attack of the Bpin ligand on the electron-poorer terminal olefin carbon of the styrene molecule (as a result of the electron-withdrawing property of the phenyl substituent) [76].

In the DFT study [76], effect of substituents with different electronic properties on the regioselectivity of the insertion reactions of olefins into Cu–B bonds in copper(I) boryl complexes has been systematically investigated. Through the systematic study, the general conclusions reached are summarized in Scheme 5. Olefins bearing an electron-withdrawing substituent have smaller insertion barriers than those bearing an electron-donating substituent. For olefins bearing an electron-withdrawing substituent, 2,1-insertion, i.e., migration of the boryl ligand to the unsubstituted carbon, is preferred over 1,2-insertion. However, the factors governing the insertion regioselectivity are more complicated for olefins bearing an electron-donating substituent as electronic and steric effects of an electron-donating substituent have opposing influence, and 2,1- and 1,2-insertions have comparable transition state energies.

4 Boryl Anions

Very recently, diamino-substituted boryl, $-B(R'NCH=CHNR')$ $(R' = 2,6\text{-}^iPr_2C_6H_3)$, complexes of Li^+, Mg^{2+}, and Zn^{2+} have been synthesized and structurally characterized [78–82]. As expected, the boryl anion displays its nucleophilicity and undergoes

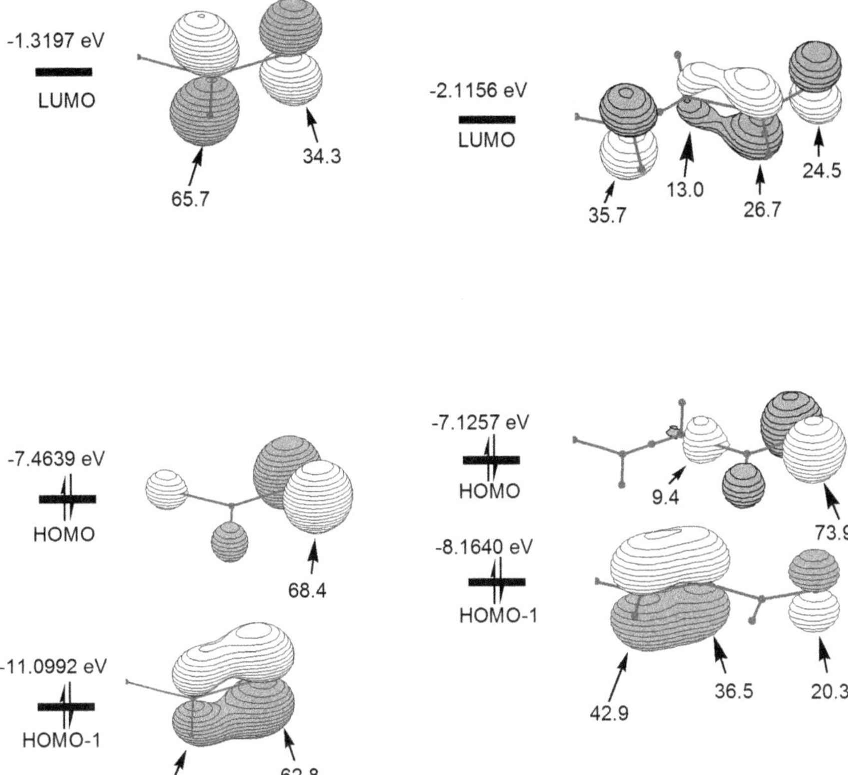

Fig. 3 Frontier molecular orbitals calculated for formaldehyde and acrolein (Adapted from [75]

Scheme 4

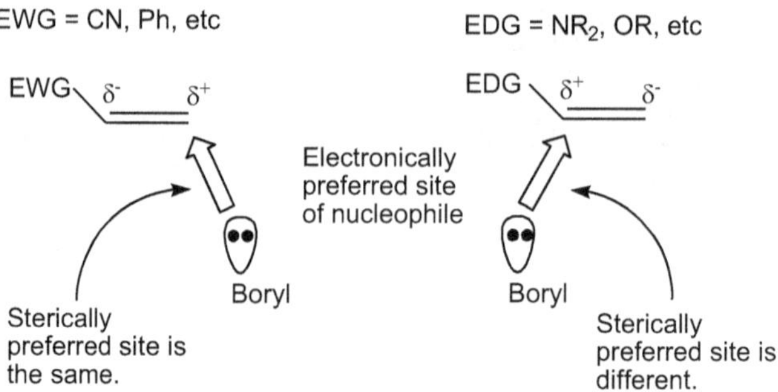

Scheme 5

nucleophilic reactions with a variety of electrophiles such as CO_2 and carbonyl compounds [80].

When the nucleophilicity is considered, one would expect that reactions of the boryl anion with organohalides RX would give R-boryl, S_N2 substitution products. However, experimentally, it was found that when the lithium boryl **27** reacted with organohalides RX, nucleophilic substitution to give RB(R′NCH=CHNR′) and/or halogen abstraction to give XB(R′NCH=CHNR′) were observed [80]. Scheme 6 shows the interesting experimental results observed for the reactions of the lithium boryl **27** with a few representative organohalides RX. In a recent article, the interesting chemoselectivity was explored in detail with the aid of density functional theory (DFT) calculations to understand how different organohalides affect the outcome of reactions with boryl anions [83].

Through the DFT study, it was found that the boryl anion can be considered as a soft nucleophile and therefore in some cases shows a preference to attack at X rather than R in an RX compound when X=Cl, Br, or I which has the ability to engage in hypervalent bonding. In all of these reactions, the boryl anion again acts as a strong nucleophile attacking the halide-bonded carbon atom to form RB (MeNCH=CHNMe) (**28**) (an expected S_N2 substitution product) and/or attacking the halogen atom to form XB(MeNCH=CHNMe) (**29**) (a halogen-abstraction product). Figure 4 shows the two pathways for the reaction of the lithium boryl with an organohalide.

The DFT results showed that the expected S_N2 substitution reactions are thermodynamically much more favorable than the halogen-abstraction reactions, indicating that the halogen-abstraction products obtained in these reactions are kinetic products. For the reactions of **27** with CH_3X, the expected S_N2 substitution pathway is kinetically favored when X=Cl while the halogen-abstraction pathway is kinetically favored when X=Br and I. For the reactions of **27** with $PhCH_2X$, the halogen-abstraction pathways are kinetically favored for all X (X=Cl, Br or I). Clearly, an organohalide having a halogen with lower electronegativity and greater ability to engage in hypervalent bonding promotes the halogen abstraction reaction. Benzyl halides were also found to promote halogen abstraction due to conjugation which stabilizes the benzyl anion in the transition state.

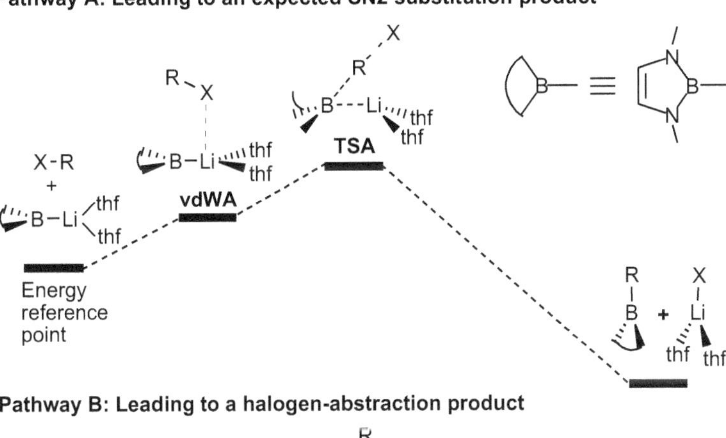

Scheme 6

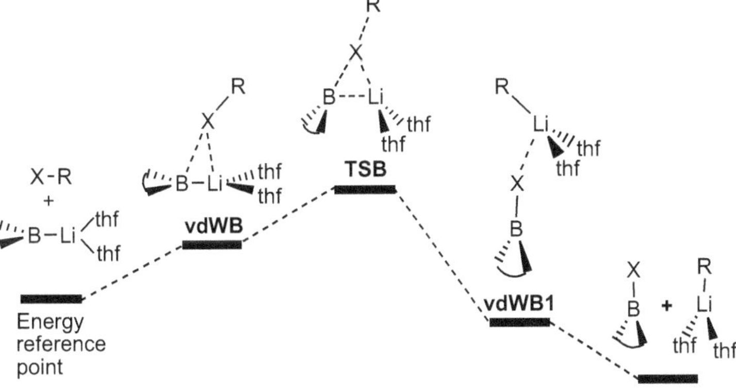

Fig. 4 Schematic reaction pathways for the reactions of the lithium boryl with organohalides (Adapted from [83])

As a side note, it is worth mentioning that NHC-stabilized π-boryl anions (**30** and **31**) were recently reported [84–86]. The potassium salt of **30** has been synthesized and structurally characterized [84]. Computational and reactivity studies indicate that there is a π-nucleophilic boron atom and **30** cleanly reacts with the electrophile MeI [84]. **31** was suggested as a reactive intermediate, which easily reacts with a diverse range of electrophiles to give different types of NHC boranes [86].

Mes = 2,4,6-trimethylphenyl   dipp = 2,6-diisopropylphenyl

30   **31**

5 Electrophilic Properties of Boryl Ligands

As mentioned above, it has been generally accepted that the π-interaction is relatively weak between a boryl ligand and a transition metal center. However, the presence of an "empty" p orbital associated with a boryl ligand can sometimes lead a boryl ligand to act as an electrophile through interaction with an electron-rich moiety, such as a ligand or a metal center.

Transition metal σ-borane complexes (**32**) containing η^2 coordinated H–B bonds have a common structural feature similar to what have been seen in many transition metal boryl complexes, i.e., the boron center in these $L_nM(\eta^2\text{-HBX}_2)$ complexes lies in a plane containing the metal center M and the two boron-bonded atoms from the two X groups [87]. Therefore, at one extreme end of the bonding continuum, we can alternatively describe the σ-borane complexes as boryl hydride complexes having a very strong bonding interaction between the hydride ligand and the electron-deficient boryl boron center. In such a bonding description, boryl ligands clearly display electrophilic properties. The structures and bonding in various transition metal σ-borane complexes reported up to 2008 have been reviewed in detail [87]. Since 2008, σ-borane complexes continue to attract considerable interest. Recent work of Alcaraz, Clot, Sabo-Etienne, and their co-workers in this field leads to synthesis, characterization, and bonding studies of a number of new and interesting σ-borane complexes (**33** and **34**) [88–95]. The work of Pandey on the bonding studies is also worth mentioning here [96, 97]. This article mainly deals with boryl ligands, and therefore, readers are suggested to refer the review article [87] and the recent work just mentioned for discussion of σ-borane complexes.

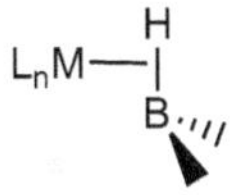

32

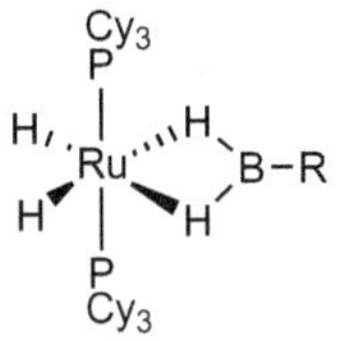

R = Mesityl, NH$_2$, N^{i}Pr$_2$, etc.

33   **34**

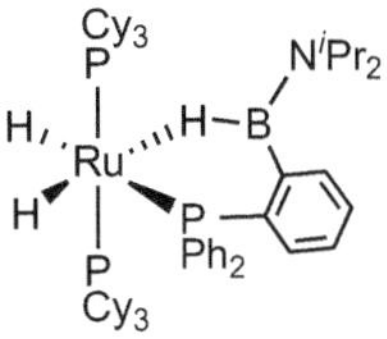

The most elegant examples to display the electrophilic properties of boryl ligands are the synthesis and characterization of the base-adducts of boryl complexes shown in Scheme 7 [98–100]. In these two examples, the iron boryl complexes each react with a ligand or an electron-rich metal fragment to form new base-adduct complexes. In the other two reported dinuclear complexes (**35** and **36**) [101, 102], each has a semi-bridging boryl ligand displaying electrophilic properties. A dative bonding interaction of an occupied d orbital from the distal metal center with the bridging boryl's "empty" p orbital has been achieved.

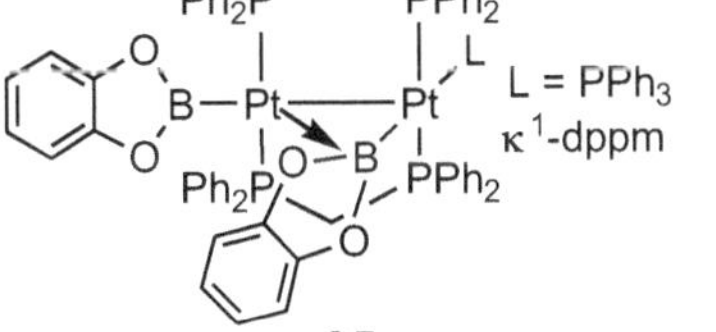

35 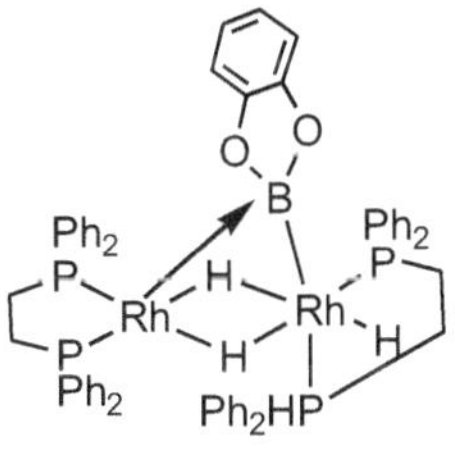   **36**

A further example of an electrophilic boryl was discussed in the C–F bond activation reaction of (PEt$_3$)$_3$Rh(Bpin) + pentafluoropyridine C$_5$NF$_5$ → (PEt$_3$)$_3$Rh (2-C$_5$NF$_4$) + FBpin [103]. The mechanism of the C–F bond activation reaction was investigated using DFT calculations. It was found that a boryl-assisted process is responsible for the C–F bond activation [103].

Recently, we studied the detailed mechanisms of the diboration of acyclic α,β-unsaturated carbonyl compounds, acrolein, methylacrylate, and dimethylfumarate (DMFU), catalyzed by a Pt(0) complex, which were experimentally

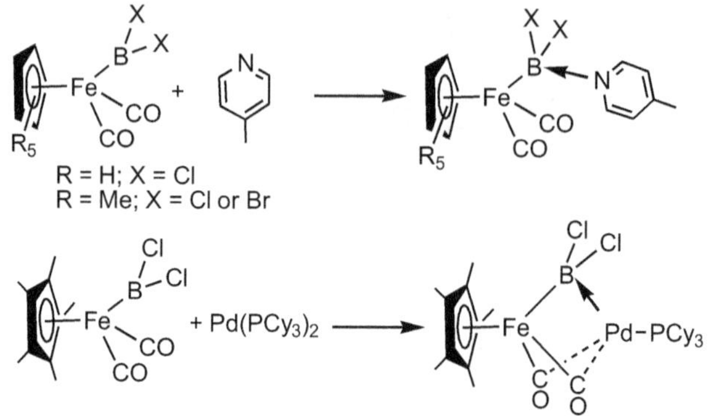

R = H; X = Cl
R = Me; X = Cl or Br

Scheme 7

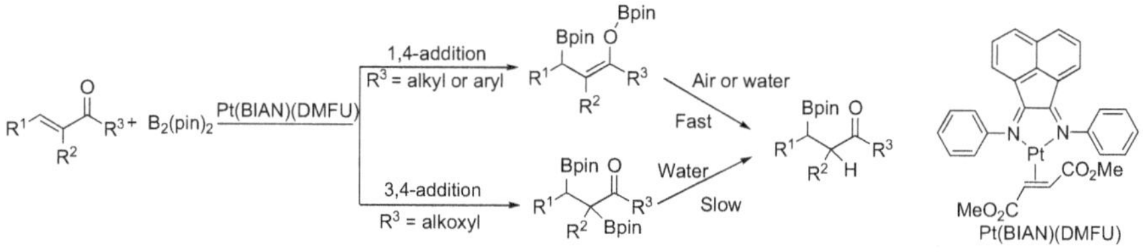

Scheme 8

reported by Norman, Marder, and co-workers (Scheme 8) [104, 105], with the aid of density functional theory calculations [106]. Through the DFT calculations, it was found that a platinum diboryl species, derived from oxidative addition of the diboron reagent to the Pt(0) complex, acts as the active species in which one of the two boryl ligands electrophilically attacks the carbonyl oxygen of a coordinated carbonyl substrate.

Let us take acrolein as an example to illustrate the findings from the DFT calculations. For a platinum boryl complex (**37**), the boryl ligand can act as a nucleophile if the Pt-boryl σ-bonding electron pair plays the role to initiate the reaction and as an electrophile if the "empty" p orbital to initiate the reaction. Scheme 9 shows three possible insertion pathways for the reaction of the bis-boryl diimine platinum(II) complex with acrolein whose π-electron density distribution is also illustrated in **37**. The free energy barriers for the pathways of the boryl attacks at C4 and C2 were calculated to be 35.7 and 53.3 kcal/mol, respectively [106]. Interestingly, the free energy barrier for the pathway of the boryl attack at O was calculated to be only 18.6 kcal/mol [106]. These results indicate that the boryl ligand, in this case, behaves as an electrophile.

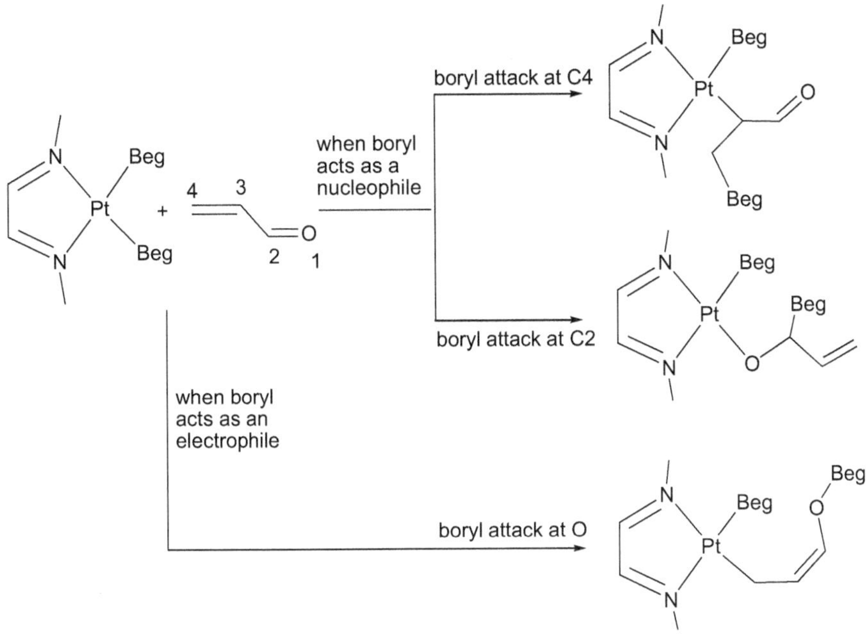

Scheme 9

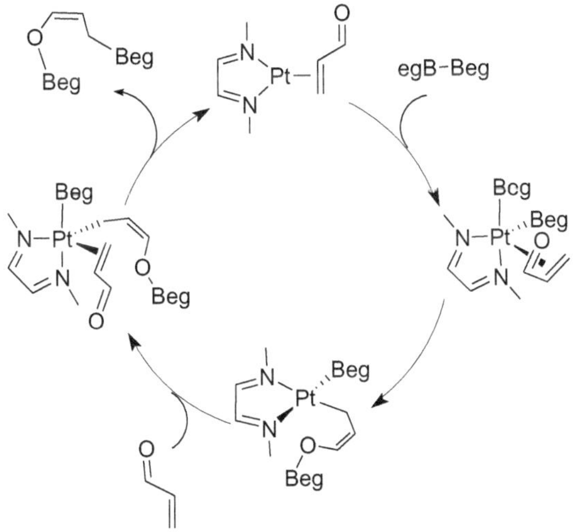

Scheme 10

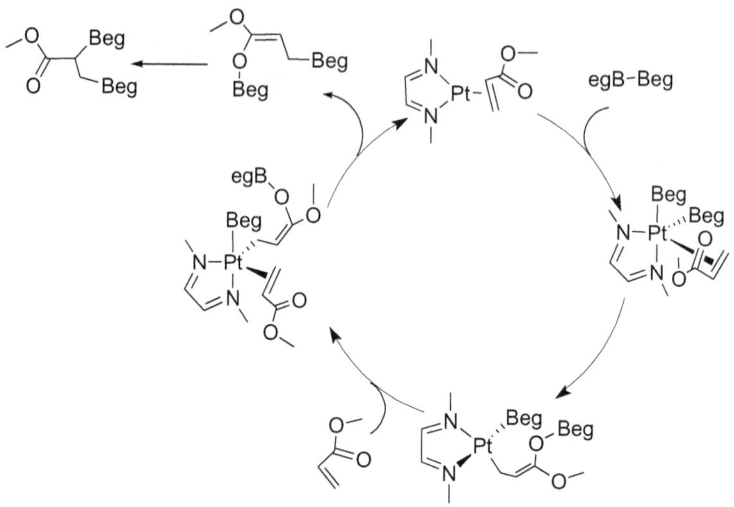

Scheme 11

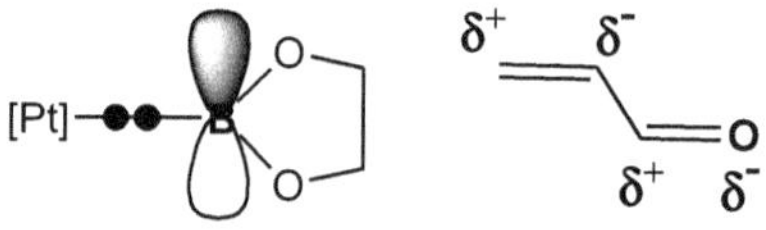

37

Through the DFT calculations, a mechanism for the Pt(0)-catalyzed borylation of acrolein was proposed (Scheme 10), which involves oxidative addition of diboron reagents, acrolein insertion into the Pt–B bond, acrolein coordination, and reductive elimination [106]. A similar mechanism has also been proposed for the Pt(0)-catalyzed borylation of methylacrylate (Scheme 11) [106].

Clearly, the platinum(II) boryls display very different chemical reactivity toward acyclic α,β-unsaturated carbonyl compounds when compared with the copper (I) boryls. The different reactivity has been explained through a bond polarization difference between a Cu–B bond and a Pt–B bond, i.e., a Cu–B bond is polarized toward B while a Pt–B bond is polarized toward Pt since the electronegativity of B is greater than Cu but smaller than Pt (Scheme 12) [106]. This bond polarization argument is also supported by the calculated NBO (natural bond orbital) atomic charges of Cu, Pt, and B atoms in the related boryl complexes (Scheme 12) [106]. It should be noted here that the M–B bond polarization was calculated for various boryl moieties and was used to understand their nucleophilic and electrophilic character [107].

initiates the reaction and
acts as a nucleophile

initiates the reaction and
acts as an electrophile

Electronegativity	Cu: 1.9 B: 2.0	Pt: 2.2 B: 2.0
Relative atomic charges	Cu: ~ +0.50 with respect to the charge of B	Pt: ~ -0.25 with respect to the charge of B

Scheme 12

6 Summary

In this chapter, the electronic properties of boryl ligands bonded to late and post-transition metal complexes have been discussed. Boryl ligands are exceptionally strong σ-donors which display strong nucleophilic character. The three-coordinate boron center in a boryl ligand also possesses an "empty" p orbital, making boryl ligands, in some cases, display Lewis acidity and oxophilicity. Clearly, the reactivity behavior of boryl ligands depends largely on the metal center to which boryl ligands are bonded.

Acknowledgments I thank my co-workers Li Dang, Wai Han Lam, King Chung Lam, Haitao Zhao, and Jun Zhu for their contribution to the research reported here. I am particularly grateful to Professor Todd B. Marder, my experimental collaborator, for his insightful comments and suggestions on the boryl chemistry discussed in this article. I would also like to thank Professor Holger Braunschweig and Dr. Rian Dewhurst for their valuable comments. I also want to acknowledge financial support from the Hong Kong Research Grants Council (Grant Nos. HKUST 603711, 603313 and CUHK7/CRF/12G).

References

1. Baker RT, Ovenall DW, Calabrese JC, Westcott SA, Taylor NJ, Williams ID, Marder TB (1990) J Am Chem Soc 112:9399
2. Knorr JR, Merola JS (1990) Organometallics 9:3008
3. Hartwig JF, Waltz KM, Muhoro CN, He X, Eisenstein O, Bosque R, Maseras F (1997) In: Siebert W (ed) Advances in boron chemistry. The Royal Society of Chemistry, Cambridge, p 373
4. Wadepohl H (1997) Angew Chem Int Ed 36:2441
5. Irvine GJ, Lesley MJG, Marder TB, Norman NC, Rice CR, Robins EG, Roper WR, Whittell GR, Wright LJ (1998) Chem Rev 98:2685
6. Braunschweig H (1998) Angew Chem Int Ed 37:1786
7. Smith MR (1999) Prog Inorg Chem 48:505

8. Braunschweig H, Colling M (2001) Coord Chem Rev 223:1
9. Coombs DL, Aldridge S (2004) Coord Chem Rev 248:535
10. Braunschweig H, Whittell G (2005) Chem Eur J 11:6128
11. Braunschweig H, Kollann C, Rais D (2006) Angew Chem Int Ed 45:5254
12. Marder TB (2008) In: Fairlamb IJS, Lynam JM (eds) Specialist periodical reports: organometallic chemistry, vol 34. Royal Society of Chemistry, Cambridge, p 46
13. Kays DL, Aldridge S (2008) Struct Bonding (Berlin) 130:29
14. Dang L, Lin ZY, Marder TB (2009) Chem Commun 3987
15. Mkhalid IA, Barnard JH, Marder TB, Murphy JM, Hartwig JF (2010) Chem Rev 110:890
16. Braunschweig H, Dewhurst RD, Schneider A (2010) Chem Rev 110:3924
17. Braunschweig H, Dörfler R, Mies J, Oechsner A (2011) Chem Eur J 17:12101
18. Cid J, Gulyás H, Carbó JJ, Fernández E (2012) Chem Soc Rev 41:3558
19. Lillo V, Bonet A, Fernández E (2009) Dalton Trans 2899
20. Rablen PR, Hartwig JF, Nolan SP (1994) J Am Chem Soc 116:4121
21. Sakaki S, Kai S, Sugimoto M (1999) Organometallics 18:4825
22. Giju KT, Bickelhaupt FM, Frenking G (2000) Inorg Chem 39:4776
23. Sivignon G, Fleurat-Lassard P, Onno JM, Volatron F (2002) Inorg Chem 41:6656
24. Dickinson AA, Willock DJ, Calder RJ, Aldridge S (2002) Organometallics 21:1146
25. Cundari TR, Zhao Y (2003) Inorg Chim Acta 345:70
26. Lam WH, Shimada S, Batsanov AS, Lin Z, Marder TB, Cowan J, Howard JAK, Mason SA, McIntyre GJ (2003) Organometallics 22:4557
27. Zhu J, Lin Z, Marder TB (2005) Inorg Chem 44:9384
28. Braunschweig H, Brenner HP, Müller A, Radacki K, Rais D, Uttinger K (2007) Chem Eur J 13:7171
29. Souza FES, Nguyen P, Marder TB, Scott AJ, Clegg W (2005) Inorg Chim Acta 358:1501
30. Westcott SA, Marder TB, Baker RT, Calabrese JC (1993) Can J Chem 71:930
31. Clegg W, Lawlor FJ, Lesley G, Marder TB, Norman NC, Orpen AG, Quayle MJ, Rice CR, Scott AJ, Souza FE (1988) J Organomet Chem 550:183
32. Lesley G, Nguyen P, Taylor NJ, Marder TB, Scott AJ, Clegg W, Norman NC (1996) Organometallics 15:5137
33. Braunschweig H, Leech R, Rais D, Radacki K, Uttinger K (2008) Organometallics 27:418
34. Braunschweig H, Gruss K, Radacki K, Uttinger K (2008) Eur J Inorg Chem 1462
35. Braunschweig H, Fuβ M, Radacki K, Uttinger K (2009) Z Anorg Allg Chem 635:208
36. Braunschweig H, Damme A, Kupfer T (2011) Angew Chem Int Ed 50:7179
37. Bauer J, Braunschweig H, Kraft K, Radacki K (2011) Angew Chem Int Ed 50:10457
38. Braunschweig H, Radacki K, Uttinger K (2008) Eur Chem J 14:7858
39. Braunschweig H, Damme A, Kupfer T (2012) Eur Chem J 18:15927
40. Arnold N, Braunschweig H, Brenner P, Jimenez-Halla JOC, Kupfer T, Radacki K (2012) Organometallics 31:1897
41. Ishiyama T, Yamamoto M, Miyaura N (1996) Chem Lett 1117
42. Iverson CN, Smith MR III (1995) J Am Chem Soc 117:4403
43. Clegg W, Johann T, Marder TB, Norman NC, Orpen AG, Peakman T, Quayle MJ, Rice CR, Scott AJ (1998) J Chem Soc Dalton Trans 1431
44. Charmant JPH, Fan C, Norman NC, Pringle PG (2007) Dalton Trans 114
45. Braunschweig H, Brenner P, Dewhurst RD, Güthlein F, Jimenez-Halla JOC, Radacki K, Wolf J, Zöllner L (2012) Chem Eur J 18:8605
46. Braunschweig H, Bertermann R, Brenner P, Burzler M, Dewhurst RD, Radacki K, Seeler F (2011) Chem Eur J 17:11828
47. Hartwig JF, Muhoro CN, He X, Eisenstein O, Bosque R, Maseras F (1996) J Am Chem Soc 118:10936
48. Lam WH, Lin Z (2000) Organometallics 19:2625
49. He X, Hartwig JF (1996) Organometallics 15:400
50. Rickard CEF, Roper WR, Williamson A, Wright LJ (2000) Organometallics 19:4344

51. Dai C, Stringer G, Marder TB, Baker RT, Scott AJ, Clegg W, Norman NC (1996) Can J Chem 74:2026
52. Clegg W, Lawlor FJ, Marder TB, Nguyen P, Norman NC, Orpen AG, Quayle MJ, Rice CR, Robins EG, Scott AJ, Souza FES, Stringer G, Whittell GR (1998) J Chem Soc Dalton Trans 301
53. Dai C, Stringer G, Marder TB, Scott AJ, Clegg W, Norman NC (1997) Inorg Chem 36:272
54. Ishiyama T, Takagi J, Ishida K, Miyaura N, Anastasi NR, Hartwig JF (2002) J Am Chem Soc 124:390
55. Lu N, Norman NC, Orpen AG, Quayle MJ, Timms PL, Whittell GR (2000) Dalton Trans 4032
56. Rickard CEF, Roper WR, Williamson A, Wright LJ (1998) Organometallics 17:4869
57. Rickard CEF, Roper WR, Williamson A, Wright LJ (1999) Angew Chem Int Ed Engl 38:1110
58. Baker RT, Calabrese JC, Westcott SA, Nguyen P, Marder TB (1993) J Am Chem Soc 115:4367
59. Marder TB, Norman NC, Rice CR, Robins EG (1997) Chem Commun 53
60. Clark GR, Irvine GJ, Roper WR, Wright LJ (2003) J Organomet Chem 680:81
61. Dai C, Stringer G, Corrigan JF, Taylor NJ, Marder TB, Norman NC (1996) J Organomet Chem 513:273
62. Lam KC, Lam WH, Lin Z, Marder TB, Norman NC (2004) Inorg Chem 43:2541
63. Rossi AR, Hoffmann R (1975) Inorg Chem 14:365
64. Takahashi K, Ishiyama T, Miyaura N (2000) Chem Lett 982
65. Takahashi K, Ishiyama T, Miyaura N (2001) J Organomet Chem 625:47
66. Laitar DS, Müller P, Sadighi JP (2005) J Am Chem Soc 127:17196
67. Zhao HT, Lin Z, Marder TB (2006) J Am Chem Soc 128:15637
68. Laitar DS, Tsui EY, Sadighi JP (2006) J Am Chem Soc 128:11036
69. Zhao HT, Dang L, Marder TB, Lin Z (2008) J Am Chem Soc 130:5586
70. Ito H, Yamanaka H, Tateiwa J, Hosomi A (2000) Tetrahedron Lett 41:6821
71. Lillo V, Prieto A, Bonet A, Díaz-Requejo MM, Ramírez J, Peŕez PJ, Fernández E (2009) Organometallics 28:659
72. Ali HA, Goldberg I, Srebnik M (2010) Organometallics 20:3962
73. Mun S, Lee JE, Yun J (2006) Org Lett 8:4887
74. Sim HS, Feng X, Yun J (2009) Chem Eur J 15:1939
75. Dang L, Lin Z, Marder TB (2008) Organometallics 27:4443
76. Dang L, Zhao HT, Lin Z, Marder TB (2007) Organometallics 26:2824
77. Laitar DS, Sadighi JP (2006) Organometallics 25:2405
78. Segawa Y, Yamashita M, Nozaki K (2006) Science 314:113
79. Kajiwara T, Terabayashi T, Yamashita M, Nozaki K (2008) Angew Chem Int Ed 47:6606
80. Segawa Y, Suzuki Y, Yamashita M, Nozaki K (2008) J Am Chem Soc 130:16069
81. Marder TB (2006) Science 314:64
82. Braunschweig H (2007) Angew Chem Int Ed 46:1946
83. Cheung MS, Mader TB, Lin Z (2011) Organometallics 30:3018
84. Braunschweig H, Chiu CW, Radacki K, Kupfer T (2010) Angew Chem Int Ed 49:2041
85. Yamashita M (2010) Angew Chem Int Ed 49:2474
86. Monot J, Solovyev A, Bonin-Dubarle H, Derat E, Curran DP, Robert M, Fensterbank L, Malacria M, Lacote E (2010) Angew Chem Int Ed 49:9166
87. Lin Z (2008) Struct Bonding (Berlin) 130:123
88. Alcaraz G, Clot E, Helmstedt U, Vendier L, Sabo-Etienne S (2007) J Am Chem Soc 129:8704
89. Alcaraz G, Grellier M, Sabo-Etienne S (2009) Acc Chem Res 42:1640
90. Alcaraz G, Vendier L, Clot E, Sabo-Etienne S (2010) Angew Chem Int Ed 49:918
91. Alcaraz G, Chaplin AB, Stevens CJ, Clot E, Vendier L, Weller S, Sabo-Etienne S (2010) Organometallics 29:5591

92. Bénac-Lestrille G, Helmstedt U, Vendier L, Alcaraz G, Clot E, Sabo-Etienne S (2011) Inorg Chem 50:11039
93. Gloaguen Y, Bénac-Lestrille G, Vendier L, Helmstedt U, Clot E, Alcaraz G, Sabo-Etienne S (2013) Organometallics 32:4868
94. Gloaguen Y, Alcaraz G, Pécharman AF, Clot E, Vendier L, Sabo-Etienne S (2009) Angew Chem Int Ed 48:2964
95. Gloaguen Y, Alcaraz G, Petit AS, Clot E, Coppel Y, Vendier L, Sabo-Etienne S (2011) J Am Chem Soc 133:17232
96. Pandey KK (2009) Coord Chem Rev 253:37
97. Pandey KK (2012) Dalton Trans 41:3278
98. Braunschweig H, Radacki K, Seeler F, Whittell GR (2004) Organometallics 23:4178
99. Braunschweig H, Radacki K, Seeler F, Whittell GR (2006) Organometallics 25:4605
100. Braunschweig H, Radacki K, Rais D, Whittell GR (2005) Angew Chem Int Ed 44:1192
101. Curtis D, Lesley, MJG, Norman NC, Orpen AG, Starbuck J (1999) J Chem Soc Dalton Trans 1687
102. Westcott SA, Marder TB, Baker RT, Harlow RL, Calabrese JC, Lam KC, Lin Z (2004) Polyhedron 23:2665
103. Teltewskoi M, Panetier JA, Macgregor SA, Braun T (2010) Angew Chem Int Ed 49:3947
104. Lawson YG, Lesley MJG, Marder TB, Norman NC, Rice CR (1997) Chem Commun 2051
105. Bell NJ, Cox AJ, Cameron NR, Evans JSO, Marder TB, Duin MA, Elsevier CJ, Baucherel X, Tulloch AAD, Tooze RP (2004) Chem Commun 1854
106. Liu B, Gao M, Dang L, Zhao HT, Marder TB, Lin Z (2012) Organometallics 31:3410
107. Cid J, Carbó JJ, Fernández E (2012) Chem Eur J 18:12794

Struct Bond (2016) 167: 59–80
DOI: 10.1007/430_2015_188
© Springer International Publishing Switzerland 2015
Published online: 10 November 2015

QM/MM Calculations on Selectivity in Homogeneous Catalysis

Jesús Jover and Feliu Maseras

Abstract The application of QM/MM methods to the study of the reaction mechanisms involved in chemo-, regio-, and enantio- selective processes has been a very productive area of research in the last two decades. This review summarizes basic general ideas in both QM/MM methods and the computational study of selectivity and presents selected results on the study of three of the most representative examples of these applications: rhodium-catalyzed hydrogenation, rhodium-catalyzed hydroformylation, and copper-catalyzed cyclopropanation.

Keywords Cyclopropanation · Hydroformylation · Hydrogenation · QM/MM calculations · Selectivity

Contents

J. Jover
Inorganic Chemistry Department, Faculty of Chemistry, University of Barcelona, c/Martí i Franquès, 1, 08028 Barcelona, Spain

F. Maseras (✉)
Institute of Chemical Research of Catalonia (ICIQ), The Barcelona Institute of Science and Technology, Avgda. Països Catalans, 16, Tarragona, 43007 Catalonia, Spain

Departament de Química, Universitat Autònoma de Barcelona, Bellaterra, 08193 Catalonia, Spain
e-mail: fmaseras@iciq.es

Abbreviations

AMPP	Aminophosphane phosphinite
B3LYP	Becke 3-parameter Lee Yang Parr
BINAPHOS	2-(Diphenylphosphino)-1,1-binaphthalene-2,2-diylphosphite
Box	Bis-oxazoline
BP	Becke Perdew
DFT	Density functional theory
DIOP	(2,3-O-isopropylidene-2,3-dihydroxy-1,4- bis(diphenylphosphino) butane)
DIPAMP	Ethane-1,2-diylbis[(2-methoxyphenyl)phenylphosphane]
DuPHOS	1,2-Bis(2,5-dimethylpholano)benzene
ee	Enantiomeric excess
er	Enantiomeric ratio
IMOMM	Integrated Molecular Orbital Molecular Mechanics
L-DOPA	L-3,4-Dihydroxyphenylalanine
MM	Molecular mechanics
ONIOM	Our own n-layered integrated molecular orbital molecular mechanics
Q2MM	Quantum to molecular mechanics
QM	Quantum mechanics
QM/MM	Quantum mechanics/molecular mechanics
QSAR	Quantitative structure activity relationships
UFF	Universal force field
Xantphos	4,5-Bis(diphenylphosphino)-9,9-dimethylxanthene

1 Introduction

The application of computational chemistry to homogeneous catalysis has made large progress in recent decades [1–3]. This progress has been driven mostly by the increase of computer power and the development of new computational algorithms. One of the significant advances from a methodological point of view has been the widespread availability of quantum mechanics/molecular mechanics (QM/MM) methods. These methods had been traditionally developed for biochemical applications [4] but have gained importance in homogeneous catalysis mostly through the expansion of the ONIOM method [5].

QM/MM methods have been a good fit in systems where an accurate QM description is mandatory for a specific region, but a large system is nevertheless required for a proper description of the reactivity. It is not surprising thus that they have been applied widely in enzyme chemistry, where the reaction is usually confined at the active center. But this prescription fits also well to transition metal chemistry, where the metal and its immediate environment often require a QM method, but the steric effects of the potentially bulky ligands can be usually described in a satisfactory way by a force field. Regioselective and enantioselective catalysts occupy a prominent place among the transition metal complexes where the description of the chemistry requires the explicit introduction of the real ligands, even if they are bulky. These bulky ligands often carry the stereogenic centers that define the selectivity of the processes.

The application of QM/MM methods to different branches of chemistry has been the subject of a number of reviews. Some of them have in particular focused on its application to inorganic systems and homogeneous catalysis [6–12]. The use of computational chemistry for the characterization of selectivity has been also the subject of other reviews [13–15]. This review intends to fill a gap because it focuses on the specific application of QM/MM methods to selectivity. After two sections dealing with a description of QM/MM methods and of computational approaches to selectivity, we will present a summary of results on three representative reactions that have been studied with these methods. The last section will collect conclusions and perspectives.

2 QM/MM Methods

QM/MM methods are a particular case of multilayer methods. Research on this topic was awarded the Nobel Prize in Chemistry in 2013. Work by Warshel, Levitt and Karplus, the three Nobel laureates, is fundamental in the formulation of the approach and its early developments [16, 17]. A lot of theoretical chemists have contributed to further development of these methods, and among them, one must cite the work by Thiel [4] and Morokuma [5, 18–20].

The basic idea behind any multilayer method is the use of different methods for the description of different regions of a chemical system. The energy description of QM/MM methods can be rigorously fit to a formula of the type:

$$E_{QM/MM}(TOT) = E_{QM}(QM) + E_{MM}(MM) + E_{QM/MM}(QM/MM). \quad (1)$$

In Eq. (1) the labels in subscript correspond to the methodological description and the labels in parentheses to the region involved. The interactions within the QM region are described at the QM level, and the interactions within the MM region are described at the MM level. The specific features of each QM/MM method are defined by the way it manages the interaction between the QM and MM regions. In principle, this interaction can be defined at either the QM or MM level, resulting in Eq. (2).

$$E_{QM/MM}(QM/MM) = E_{QM}(QM/MM)) + E_{MM}(QM/MM). \quad (2)$$

The MM description of this interaction has been defined as mechanical embedding and the QM description as electronic embedding [21]. The mechanical embedding can be roughly assigned to the qualitative concept of steric effects, with the electronic embedding corresponding to electronic effects. Mechanical embedding is easier to implement in a computational code than electronic embedding, and because of this, pure mechanical embedding may be considered as zero order QM/MM.

A method with only mechanical embedding would simplify the formula above to the following expression:

$$E_{QM/MM}(TOT) = E_{QM}(QM) + E_{MM}(MM) + E_{MM}(QM/MM). \tag{3}$$

Inspection of Eq. (3) shows that it would correspond to the total MM energy by changing the subscript in the first term. This leads to the subtractive scheme characteristic of the ONIOM method [5, 18–20]:

$$E_{QM/MM}(TOT) = E_{QM}(QM) + E_{MM}(TOT) - E_{MM}(QM). \tag{4}$$

The formula in Eq. (4) represented an advance in QM/MM methodology because it defined an approach to compute QM/MM energy relying mostly in previously available QM and MM codes, with a minimum interface to be defined between them. In contrast, previous QM/MM implementations had required the explicit introduction of additional terms in the energy expression.

The use of Eq. (4) defines a path to compute the QM/MM energy and its derivatives for any system, and most of the results discussed in this review are based in this formula. There are however a few caveats about the limitations of this formula and its practical implementation that deserve comment.

The first issue is the handling of electronic embedding, which is neglected in the expression in Eq. (4). This is a limitation that can be corrected by the introduction of additional terms in the formula, and this has been indeed implemented also within the ONIOM methodology [22]. Electronic embedding is usually introduced as electrostatic embedding, by placing point charges affecting the QM energy on the positions where the MM atoms are placed. The particular value of these point charges has been heavily discussed in the literature [21]. On the other hand, the neglect of electronic effects may be viewed as an advantage for the analysis of the results, as it provides an opportunity to compute separately electronic and steric contributions [23]. Pure steric contributions can be computed from the QM/MM calculation. The full QM calculation provides both steric and electronic effects, and the latter can be extracted by comparison with the QM/MM calculation.

The second issue is the handling of connections between the QM and MM regions. If there is a covalent bond across the QM/MM partition, the simple deletion of the MM region would leave a dangling bond in the QM calculation. This QM calculation would not be a realistic model of the QM region in the real system. Two main solutions have been given to this problem, special orbitals or link atoms. The introduction of special orbitals in the border atoms in the QM region is the more sophisticated approach, making easier the transfer of electronic effects from the MM region to the QM region [24, 25]. The results presented in this contribution will not use these special orbitals, which are therefore not discussed here in detail.

The alternative approach, used in the reported results, is the introduction of the link atoms [18]. These are extra atoms, usually hydrogen atoms, which are introduced to cap the dangling bonds in the QM regions. In order to avoid the introduction of artificial degrees of freedom in the system, the placement of these link atoms is completely defined by that of the corresponding atoms in the real system in modern QM/MM implementations.

In summary, there is a large variety of QM/MM methods. Most applications in selectivity problems in transition metal chemistry have been however concentrated in the use of the IMOMM and ONIOM implementations with pure mechanical embedding, IMOMM being an earlier version of ONIOM. These are simple implementations that allow the introduction of steric effects at low computational cost and with moderate effort in the construction of the input files.

3 Computing Selectivity

Selectivity is an important concept in chemistry [26]. It can be defined as the preferential outcome of a chemical process over a set of other plausible outcomes. A typical example is the case of enantioselectivity, where two enantiomers are in principle possible, but one of them is preferentially obtained. Enantioselectivity is one of the most interesting and most extreme cases of selectivity, as enantiomeric products have by definition the same energy. But the concept of selectivity is more general. Regioselectivity refers to the preference of one direction or chemical bond making or breaking above other alternative directions. Chemoselectivity refers to the reactivity of a chemical functional group in the presence of others.

The concept of selectivity is often associated with organic chemistry, with its clear definition of stereocenters and functional groups. It can be in general applied to either stoichiometric or catalytic processes. The case of catalytic processes is particularly intriguing, as it is possible to introduce a catalyst that induces the selective formation of only one of the possible products. The species that induces selectivity is often more difficult to synthesize than the reactants, and because of this, its introduction as a low-concentration catalyst is desirable. Despite the recent progress in organocatalysis, transition metal catalysis plays still a key role in homogeneous catalysis. Because of this, the prediction of selectivity has been an important goal for homogeneous catalysis and the application of QM/MM methods to the field.

The translation from the typical free energy profiles obtained from pure QM or QM/MM calculations to selectivity ratio or selectivity excess requires a straightforward application of formal kinetics [27] that will be briefly described in this section.

Let us assume that a reactant A can evolve to two different products B, C through first-order kinetics ruled by rate constants k_b, k_c.

$$\frac{d[\text{B}]}{dt} = k_{\text{b}}[\text{A}] \quad \frac{d[\text{C}]}{dt} = k_{\text{c}}[\text{A}] \, . \tag{5}$$

Dividing the two formulas in Eq. (5), integrating and assuming an initial concentration of zero for both B and C, we reach the expression

$$\frac{[\text{B}]}{[\text{C}]} = \frac{k_{\text{b}}}{k_{\text{c}}} \, . \tag{6}$$

Equation (6) states a constant ratio between the concentrations of B and C throughout time, which is defined by the ratio between the rate constants. The rate constants can be related through the Eyring equation to the activation free energies.

$$k_{\text{b}} = \frac{k_{\text{B}}T}{h} \exp\left(\frac{-\Delta G_{\text{b}}^{\ddagger}}{RT}\right) \quad k_{\text{c}} = \frac{k_{\text{B}}T}{h} \exp\left(\frac{-\Delta G_{\text{c}}^{\ddagger}}{RT}\right) . \tag{7}$$

Inserting the Eyring equations into Eq. (6), we obtain the relationship between the concentrations and the activation free energies.

$$\frac{[\text{B}]}{[\text{C}]} = \exp\left(\frac{\Delta G_{\text{c}}^{\ddagger} - \Delta G_{\text{b}}^{\ddagger}}{RT}\right) . \tag{8}$$

The expression in Eq. (8) is strictly valid only for cases where first-order kinetics are present, but it is easy to generalize to other cases. The key to its simplicity is that the paths leading to either B or C start from a common intermediate, but this is one of the usual features of chemical problems where selectivity is considered. Extra species leading to a second-order expression could be introduced in the expression in a straightforward way. The usual complication of multistep processes can be properly handled by using the energy span model [28], which simplifies the kinetic expression.

Equation (8) gives the selectivity ratio. When enantioselectivity is considered, the enantiomeric excess (*ee*) is usually provided. It can be obtained from the enantioselectivity ratio (*er*) through the following formula:

$$ee = \frac{er - 1}{er + 1} \, . \tag{9}$$

The exponential dependence of selectivity ratio (or selectivity excess) with respect to the free energy differences underlines one of the key features of selectivity in chemistry: its strong sensitivity to small energy changes. A free energy difference in the barriers of 1 kcal mol^{-1} at a temperature of 298 K translates roughly into a selectivity ratio of 5:1 (84:16), a difference of 2 kcal mol^{-1} leads to a ratio of 29:1 (97:3), and a difference of 3 kcal mol^{-1} leads to a ratio of 153:1 (99:1). A difference of 5 kcal mol^{-1} would already lead to the non-observation of the minor product, which should only be present in the range of the 0.02%.

The strong sensitivity of selectivity to differences in free energy barriers poses a challenge to computational chemistry. Practical reactions are often considered to be efficient when selectivity is above 90%, and this boils down to differences between transition states inferior to 2 kcal mol^{-1}. This level of absolute accuracy is difficult to achieve for most modern computational techniques on realistic chemical models. Luckily there is a phenomenon of cancelation of errors which very often favors selectivity calculations. It is not critical to have an error well below 1 kcal mol^{-1} in the absolute free energy barriers but in the difference in free energy barriers leading to the different products. And it so happens that most of the absolute error in the computed barriers is nearly identical in all the competing paths. This can be better understood with an example on rhodium-catalyzed enantioselective hydrogenation that will be discussed in detail in the following sections. The absolute barrier to the key transition state depends on the interactions between substrate and metal that are difficult to reproduce exactly. But the nature of these interactions is the same in the transition states leading to the R and S products. The discrimination between the two transition states comes from the steric interactions between the substituents, and these purely steric interactions are more likely to be well reproduced if the force field is sufficiently accurate.

A closely related issue is that the difference of activation free energies has been often replaced in practice by the difference in activation enthalpies or activation potential energies. The validity of this approach is again related to cancelation of errors. The correction necessary to convert the potential energy barrier to free energy barrier for one product is often nearly the same to the correction required to convert the potential energy barrier leading to the alternative product.

4 Rhodium-Catalyzed Hydrogenation of Enamides

The first homogeneous hydrogenation of alkenes, using $RuCl_3$, was reported in the early 1960s by Halpern, Harrod, and James [29]. Later on, a similar reaction was published by Wilkinson using his famous [$RhCl(PPh_3)_3$] catalyst [30]. These processes, combined to the contemporary development of homogeneous asymmetric reactions by Noyori and Nozaki [31], gave birth to the asymmetric hydrogenation of olefins first reported by Knowles and Horner in 1968 [32]. This reaction employed monodentate phosphane ligands (L*) as sources of chirality and provided poor enantiomeric excesses (Fig. 1).

The usage of more complex ligands, e.g., DIOP [33], led to the improvement of these reactions and allowed the higher *ee*'s to be obtained. One important application of this methodology was the industrial synthesis of L-DOPA, developed in the group of Knowles, employing the bidentate ligand DIPAMP.

The mechanism of this reaction, fully supported with experimental data, is shown in Fig. 2 [34]; this pathway is known as the alkene or the unsaturated

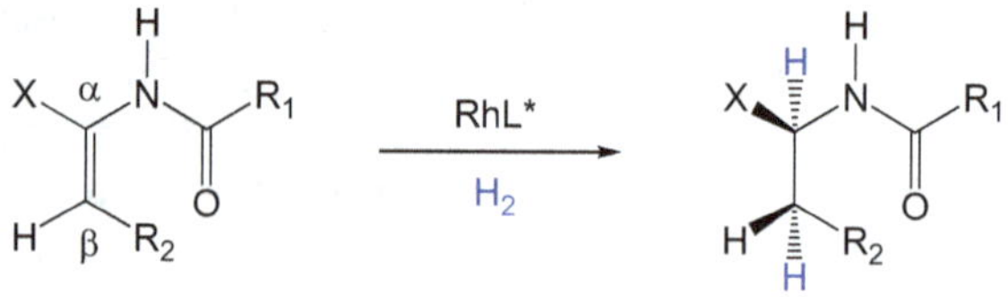

Fig. 1 First reported asymmetric rhodium-catalyzed hydrogenation of olefins

mechanism. The catalytic cycle starts by the replacement of the two solvent molecules (S) on the initial rhodium (I) complex by the enamide substrate. Then the hydrogen comes in, and the oxidative addition takes place, delivering the Rh (III) dihydride intermediate. After that, one of the hydrides is transferred either to the α or β position of the olefin (Fig. 2 shows the transfer on the latter). Finally the hydrogenated product is obtained by a reductive elimination process. An alternative pathway, called the dihydride mechanism, is also plausible and has been found to be operative in reactions involving highly electron-rich P-ligands [35]. In this case the oxidative addition of the hydrogen happens before the solvent is substituted by the enamide substrate.

The first computational studies dealing with the hydrogenation of olefins are among the first reports of computed full catalytic cycles. They date from the late 1980s and were published by Morokuma and coworkers. The full mechanism of the rhodium-catalyzed hydrogenation of ethylene with the Wilkinson catalyst was reported for a model system [36]. A decade later Landis and Feldgus reported a computational study on the hydrogenation of enamides [37]. In order to determine the turnover-limiting step, they employed a small model system related to the one shown in Fig. 1: $L^*=PH_3$, $R_1=R_2=H$, and $X=CN$. Four different pathways (A–D, Fig. 3) were found to describe the approach of H_2 to the catalyst/enamide complex. In paths A and C, the H_2 moiety comes in parallel to the Rh–olefin bond, while in B and D, it is parallel to the Rh–$O_{carbonyl}$ bond. In all cases the insertion process was computed for both possible reaction centers (Cα and Cβ). For pathways A and B, the preferred insertion position is Cα, whereas Cβ is favored in pathways C and D. The lowest energy pathway was found to be A with the highest step corresponding to the insertion process. Although the model system employed allowed a quite complete study of the whole processes, it could not be used to study the enantioselectivity of this hydrogenation reaction. This was fixed by the same authors in subsequent publications where they included a full chiral diphosphine as DuPHOS in QM/MM ONIOM studies [38, 39]. The system was described with three different layers, as shown in Fig. 4, and the enamide substrate was α-formamidoacrylonitrile, the same employed previously. In this case, and depending on the orientation adopted by the substrate when coordinating the metal center, the number of studied pathways is doubled: A, B, C, and D for the pro-*S* and pro-*R* manifolds. Pathway A was found to be the most favored one, with the H_2 oxidative addition step bearing the highest energy barrier.

These calculations were able to reproduce the anti "lock-and-key" mechanism observed experimentally; the most stable pro-*S* intermediate displays the highest reaction barrier, while the pro-*R* complex, higher in energy, has the lowest barrier to

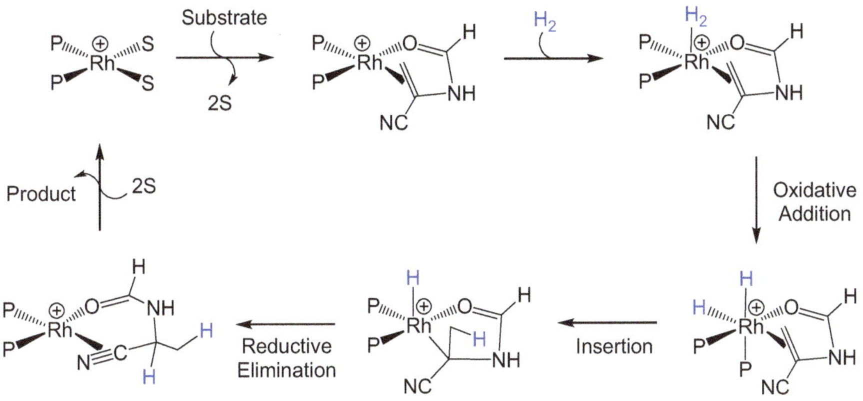

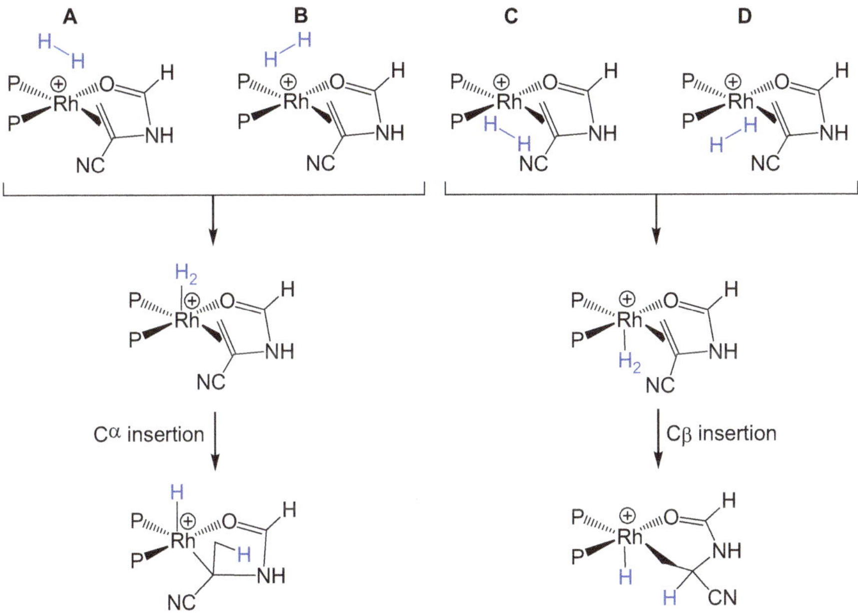

Fig. 2 Proposed mechanism for the rhodium-catalyzed hydrogenation of enamides

Fig. 3 Different approaches of H₂ (pathways A–D) to the substrate–catalyst complex in the Rh-catalyzed hydrogenation of enamides and mechanism for the insertion of Cα and Cβ into the Rh–H bond

produce the final *R* product, which is the major outcome of the reaction (Fig. 5). The energy difference between the computed transition states (4.4 kcal mol^{-1}) corresponds to a theoretical enantiomeric excess of 99.9%, well within the range obtained in experiments [40].

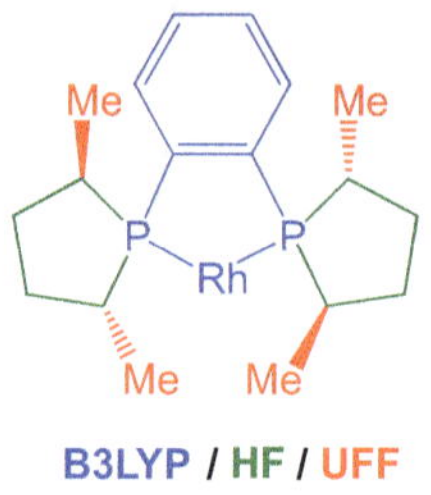

Fig. 4 Partition scheme describing the [Rh((R,R)-Me-DuPHOS)]$^+$ complex

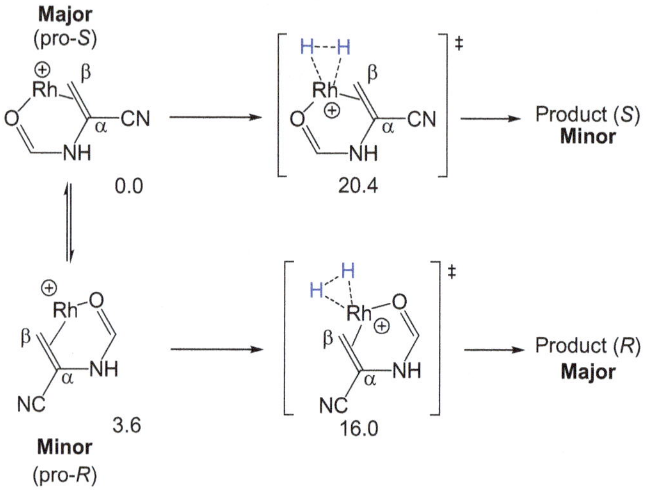

Fig. 5 Pro-*R* and pro-*S* Rh-complexes and their rate-determining oxidative addition transition states in the asymmetric hydrogenation of α-formamidoacrylonitrile, relative energies in kcal mol^{-1}

These results could be explained by using a quadrant map of the active catalyst, as originally proposed by Knowles and coworkers (Fig. 6) [41]. The pro-*S* enamide complex is more stable because Cα can be located in the Rh–phosphane plane, while in the pro-*R* analog this is not possible, and thus Cβ occupies the place in the metal–ligand plane. In contrast, when H$_2$ enters the coordination sphere of the metal, in the pro-*S* isomer the CN group is forced to move into a hindered space, which does not happen with the pro-*R* complex. Therefore, the hydrogen activation is more favorable in the latter case, and the *R* product is formed much faster. Morokuma and coworkers found similar results using the IMOMM methodology for a related system [42].

Subsequent publications pointed out to the fact that A is not always the most favorable pathway and C could arise as the preferred one depending on the electronic features of the substrate. This was first reported separately by Landis and Feldgus using QM/MM calculations [43] and by Wiest and coworkers using

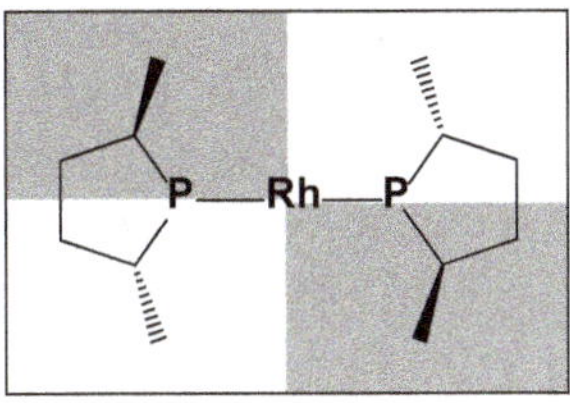

Fig. 6 Quadrant map of the Rh-diphosphine catalyst

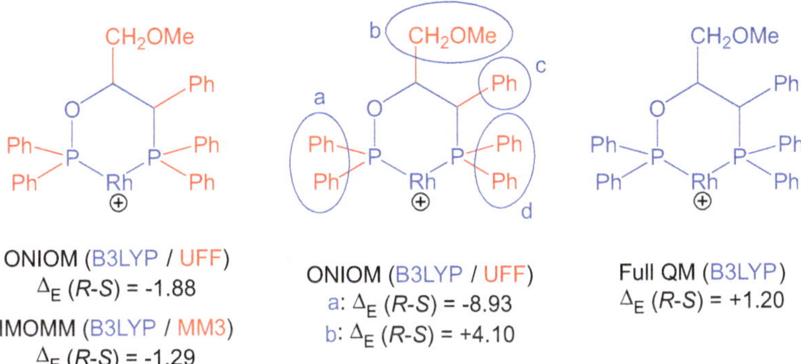

Fig. 7 QM/MM partitions and methods used for modeling the rhodium active species and the corresponding difference in energy (in kcal mol^{-1}) between the lowest Pro-S and Pro-R transition states in the hydrogenation of methyl-(N)-acetylaminoacrylate

full QM calculations [44]. In both cases it was demonstrated that pathway A is preferred whenever electron-withdrawing groups are present in Cα, favoring the insertion process on this position. In contrast, electron-donating groups can reverse this tendency and end up favoring the insertion of Cβ, thus following pathway C. This means that the R/S selectivity of the final product can be drastically changed depending on the properties of the substrate.

In all these reports, where the ligands have C$_2$ or C$_s$ symmetry, the results obtained with QM/MM or full QM calculations provide very similar results. However, Maseras and coworkers found a modification of this trend when using the bidentate phosphane–phosphinite ligand depicted in Fig. 7 in the rhodium-catalyzed hydrogenation of methyl-(N)-acetylaminoacrylate [45], with which they obtained different ee's depending on the QM/MM partition employed. When the **a–d** groups were included in the UFF region, the energy difference between the lowest transition states delivering the R and S products was found to be -1.88 kcal mol^{-1}, contradicting the experimental results where a 40% ee in S was obtained. The problem remained when the methodology was changed to IMOMM (B3LYP/MM3) calculations. The only way to obtain the right enantiomeric excess was to carry out the calculations with a full QM method, using the B3LYP functional. Single point calculations using different QM/MM partitions were employed to determine the impact of the electronic effect of the ligand substituents on the enantioselectivity. When the phenyl rings of the **a** region are included in the QM part, the energy difference in favor of the R product is strongly enhanced ($\Delta_E(R-S)=-8.93$ kcal mol^{-1}). If the **d** phenyl groups are placed within the QM region, the other enantiomer is obtained ($\Delta_E(R-S)=+4.10$ kcal mol^{-1}). These results indicate that increasing the basicity on either P-donor atom enhances the formation of one of the two different enantiomers, with the phosphinite favoring the R product and the phosphane delivering the S product. These observations are in line with what is observed experimentally and theoretically with an (S)-BINOL-derived phosphite ligand, where a 99% of ee in R was obtained [46]. Full QM calculations on this system indicate that the enantioselectivity depends both on electronic and steric effects. Four pathways are plausible in this reaction, but two of them, the ones where the olefin is *trans* to the phosphite, are avoided by electronic effects. One of the two remaining pathways is sterically blocked by the BINOL ligand, leaving just one pathway for the reaction to occur. These results explain the high selectivity found in this reaction.

QM/MM and full QM methodologies have been demonstrated to be accurate enough to study the enantioselectivities of specific asymmetric hydrogenation reactions and have helped to unravel the mechanistic complexity of these processes. These methods are, however, still quite expensive to carry out a complete screening of a real catalytic system where many combinations leading to different products may be possible. Other approaches such as QSAR-like approaches based on descriptors [47] and pure MM calculations with a specifically tailored Q2MM force field have been also productive alternatives [48, 49].

5   Rhodium-Catalyzed Hydroformylation

The hydroformylation reaction is the transition metal-mediated addition of carbon monoxide and dihydrogen to the double bond of an alkene (Fig. 8). It is one of the most important reactions catalyzed by homogeneous transition metal complexes in the industrial production of bulk chemicals [50, 51]. Figure 8 highlights the regioselectivity issue in hydroformylation: two different products are possible when the alkene is not symmetrically substituted. A common case is that where the alkene has a single substituent and two products are possible, one linear and one branched. The general mechanism of the rhodium-catalyzed hydroformylation and the particular issue of regioselectivity have been examined in a number of theoretical studies with different computational techniques and reviewed quite recently [52]. Significant progress has been made in the mechanistic understanding, but the topic has not been fully solved for all cases, and it continues to be a matter of study for pure DFT methods [53, 54]. Some significant contributions have been made from the DFT/MM point of view and are reviewed below.

Carbó et al. used a QM/MM IMOMM method [18] to study the origin of regioselectivity in the Rh-diphosphine-catalyzed hydroformylation [55]. The

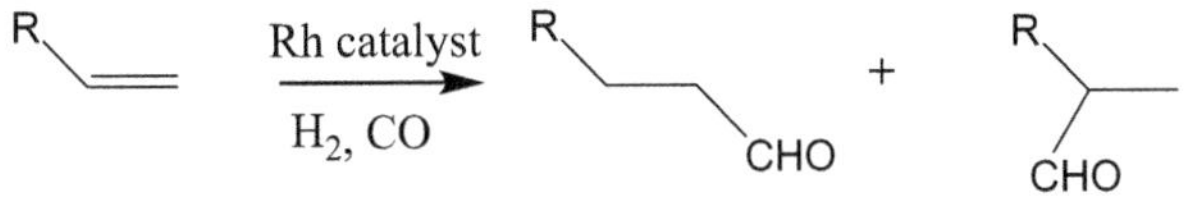

Fig. 8 The rhodium-catalyzed hydroformylation of alkenes

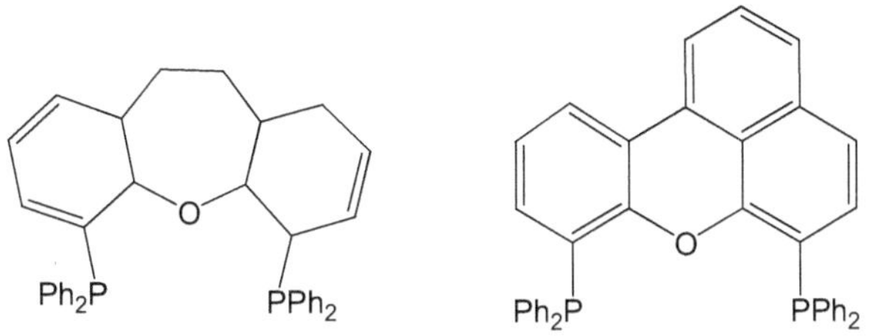

Fig. 9 The homoxantphos and benzoxantphos diphosphine ligands

study was centered on the rationalization of the experimentally observed dependence of selectivity with the bite angle of a variety of xantphos ligands. It had been observed that ligands with the larger bite angles produced a higher amount of linear product. Because of this, the key transition state in the process, corresponding to olefin insertion into the rhodium–hydride bond, was computed for two different diphosphine ligands, benzoxantphos and homoxantphos (see Fig. 9) with extreme cases of natural bite angle, and two different alkenes, propene and styrene. Only the results with propene will be discussed here.

A total of eight transition states were computed for each system, four leading to the linear product and four leading to the branched product. The results reproduced satisfactorily the experimental trends. Benzoxantphos produced bite angles around $110°$, while the bite angles for homoxantphos were around $100°$. The pattern in the relative energy distribution of transition states was the same for both ligands, but the relative energy barriers were different. The computed percentages of linear product for propene were 83% for benzoxantphos vs 73% for homoxantphos systems. This must be compared with experimental values for 1-octene of 98.1% and 89.5%, respectively. Agreement was not perfect, but the trend was correctly reproduced.

Calculations allowed moreover to perform a computational experiment where each phenyl substituent was replaced by hydrogen while maintaining the backbone of diphosphine ligands (PH_2 model). By removing the phenyl substituents, the non-bonding effects of the phenyls on regioselectivity where put aside. The PH_2 model produced significantly lower selectivities than the full model, with the values going down from 83%/73% to 74%/63%. There is an intrinsic selectivity associated with the backbone, but its effect on the overall regioselectivity is exercised through the steric interactions of the phenyls. It is not an orbital effect associated with changes in the electronic properties of the rhodium center, but a steric effect that plays through the substituents at the phosphane. Wider bite angles increase the steric interaction of the diphosphine substituents with branched species, which become more destabilized. One could therefore expect that bulkier groups than phenyl groups would lead to higher linear/branched regioselectivity ratios.

Related systems were reexamined by Zuidema et al. [56] a few years later to analyze the rate-determining step in the hydroformylation of 1-octene, catalyzed by the rhodium–xantphos catalyst system using a combination of experimentally determined kinetic isotope effects and computational approaches. The focus of this work was to clarify whether alkene coordination or hydride migration is the rate-determining step. Both ONIOM(B3LYP:UFF) and B3LYP calculations were carried out on the key catalytic steps, using the real ligand systems. The calculations quantitatively reproduced the energy barrier for CO dissociation. The overall barrier for hydride migration from the resting state was found to be 3.8 kcal mol^{-1} higher than the barrier for CO dissociation. This fits well with the experimentally determined trend, confirming the assumptions of the previous work on regioselectivity. The combination of kinetic isotope effects and theoretical studies suggested that the overall barrier for hydride migration, starting from the resting state of the catalyst, determines the activity in the rhodium–xantphos-catalyzed hydroformylation of 1-octene.

Apart from the issues concerning regioselectivity, there is also the potential for enantioselectivity, as a stereogenic center can be generated in branched products. Carbó et al. [57] applied the IMOMM (BP:Sybyl) method to analyze the origin of stereoinduction in the hydroformylation of styrene by rhodium complexes bearing chiral aminophosphane phosphinite (AMPP, see Fig. 10) ligands. The roles of the stereogenic center at the aminophosphane phosphorus atom (NP*) and of the chirality of the backbone were considered in three experimentally reported cases: (1) P-stereogenic yielding high *ee*, (2) P-nonstereogenic (R=Ph) yielding low *ee*, and (3) P-stereogenic yielding low *ee*. Experimentally observed trends for the three studied AMPP ligands were fairly well reproduced, and the calculations revealed that the different non-bonding weak-type interactions of styrene with the substituents of the NP* stereogenic center in an axial position was responsible for stereodifferentiation.

Stereoselectivity issues were also analyzed with the help of IMOMM(BP86: Sybil) calculations by Aguado-Ullate et al. [58] in the case of the reaction of styrene where the rhodium catalyst is carrying an unsymmetric bidentate phosphane phosphite ligand such as BINAPHOS. The behavior of the [Rh{(*R,S*)-BINAPHOS}(CO)$_2$H] catalyst was studied. The placement of the phosphane moiety in the apical site and the phosphite moiety on the equatorial site was shown to be critical for high enantioselectivity. The axial chirality of the phosphite discriminated one of the competitive equatorial-apical paths, whereas the chirality of the backbone discriminated one of the two enantiomers. QM/MM calculations were also used in the definition of QSAR descriptors for hydroformylation by the same group [59].

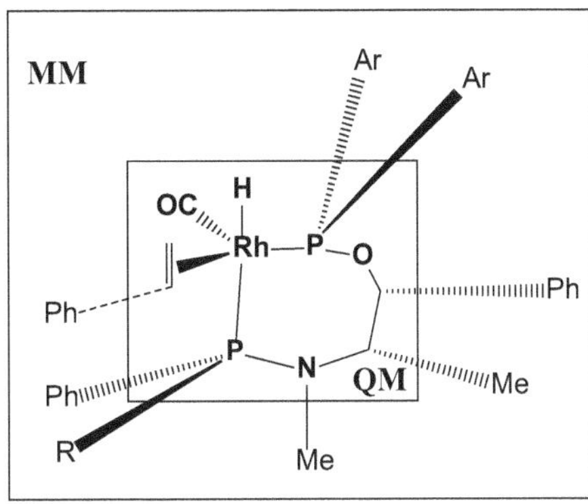

Fig. 10 QM/MM partition in the AMPP-Rh-styrene complex

6 Copper-Catalyzed Cyclopropanation

The first asymmetric copper-catalyzed cyclopropanation using a homogeneous catalyst was reported by Noyori in 1966 [31]. This reaction, which allowed the cycloaddition on styrene, was carried out with a chiral Schiff base copper complex and produced poor enantiomeric excess (Fig. 11). Further refinement of the chiral ligands produced later much better catalysts, such as the bis-oxazoline (Box) derivatives, able to provide enantioselectivities up to 99%.

Not much is experimentally known about the operating mechanism of this cyclopropanation reaction; the available data indicates that the metal remains as Cu (I) throughout the whole process and that the reaction proceeds through a metal–carbene intermediate [60]. Based on the reactivity observed for analogous catalysts, e.g., Ni(0) complexes or Zn(II) carbenoid systems, two different mechanistic pathways were suggested. The first one involves the stepwise formation of metallacyclobutane intermediates, while the alternative mechanism consists of the concerted direct carbon insertion (Fig. 12). A computational study, using DFT calculations, was employed to determine the preferred catalytic cycle for the cyclopropanation reaction of methyl diazoacetate with ethylene and a Cu(I)(Box) complex [61]. At first, a small model system, where the Box ligand was N,N'-dimethylmalonaldiimine, was used to evaluate all the reaction steps. The results stated that the rate-limiting step was the nitrogen extrusion delivering the copper–carbene intermediate, and the computed barriers indicate that the carbene insertion mechanism was preferred over the formation of the metallacycle, with energy requirements of 9.8 and 12.8 kcal mol^{-1}, respectively. Nevertheless, the stereoselectivity control comes from the reaction between the alkene and the carbene moieties. In order to check the performance of the computed mechanism with the experimental observations, these results were extended then to a more complex chiral system, where the simple starting diimine was replaced by 2,2′-methylenebis[(4S)-methyl-2-oxazoline]. Two different diastereomeric transition states were computed, corresponding to the approach of the alkene to the Si and Re faces of the plane formed by the copper–carbene complex, and the former was favored by more than 1 kcal mol^{-1} when the solvent effects were taken

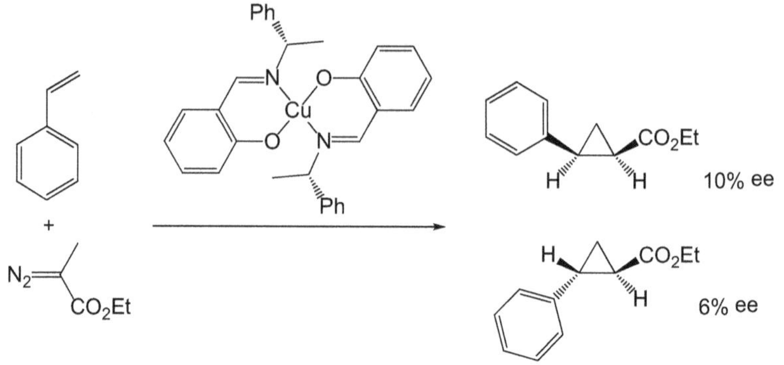

Fig. 11 First reported asymmetric copper-catalyzed cyclopropanation

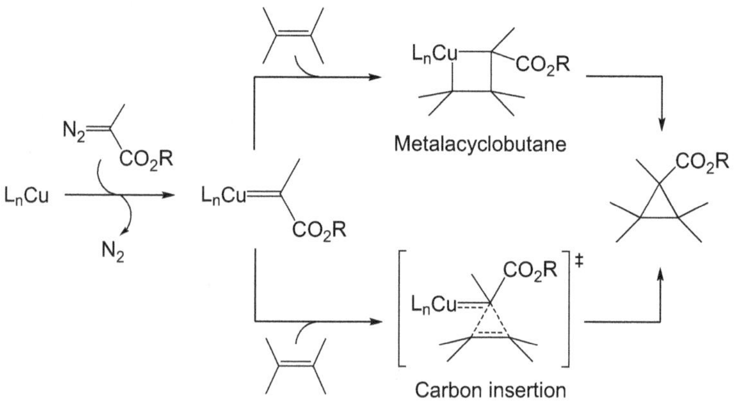

Fig. 12 Proposed mechanisms for the copper-catalyzed cyclopropanation of alkenes

into account. This behavior can be related to the steric hindrance that appears between the methyl substituent on the ligand and the ester group when the olefin approaches the copper species through the *Si* face. The computed energy difference between both transition states showed a good agreement with the enantiomeric excess obtained in the cyclopropanation reaction between ethyl diazoacetate and styrene with a similar Box ligand [62].

The *cis/trans* selectivity of this reaction was studied replacing the ethylene by 1-propene. Although eight different possible transition states arise from this modification, only in four of them the incoming methyl group of the propene lies far from the Box ligand. Thus, four transition states were computed, *Re-cis*, *Si-cis*, *Re-trans*, and *Si-trans*, and found to have energy barriers of 12.5, 13.4, 11.3, and 12.2 kcal mol^{-1}, respectively, in a nice qualitative agreement with the experiments [62]. In this case the *cis/trans* selectivity seems to be governed by the steric interaction between the substituent on the olefin and the ester group of the carbene.

The enantioselectivity of these cycloaddition reactions is strongly affected by the anion coordinated to the initial Cu(I)(Box) catalyst, e.g., replacing a triflate by a chloride in the cyclopropanation between ethyldiazoacetate and styrene produces low yields and enantioselectivities of 8 and 3% for the *cis* and *trans* products, respectively, while the triflate systems deliver 92 and 94% [63]. This counterion effect was also studied computationally by the group of García and coworkers employing the same methodology as above [64]. The counterion replacement in the small model system produces different geometries for the intermediates and transition states, and thus the relative energy differences along the reaction pathway change substantially; for instance, the barrier of the insertion stage is raised by more than 3 kcal mol^{-1}. In addition, the formation of chloride-bridged dimers seems quite possible, indicating that the whole process should be slower and suffering a remarkable decrease in the catalytic activity. The enantioselectivity reduction observed for the chlorinated system was attributed to the lower steric repulsion between the methyl group of the ligand and the incoming ester substrate in the concerted transition state. The results obtained were clearly in line with the decrease in enantioselectivity observed when chloride is employed as a counterion.

The replacement of ethylene by styrene as a more nucleophilic olefin substrate in the cyclopropanation reactions produced a major issue: the concerted transition state could not be located by classical potential energy surface explorations as reported for the first time by Norrby and coworkers [65]. In this case they studied the reaction between substituted styrenes and ethyldiazoacetate with QM/MM IMOMM using the catalyst shown in Fig. 13a. The concerted transition state was not possible to find exploring the potential energy surface because the approach of the double bond to the carbene produced a monotonic downhill energy profile. Thus, the Gibbs free energy surface was computed using a linear transit scan by optimizing a set of structures with fixed distances between the centroid of the alkene and the carbene carbon (d_1 in Fig. 13a). In this way the free energy reached maximum value at d_1 around 2.5 Å; this "transition state" is strongly asynchronous since the carbene–CHPh distance is around 2.75 Å, while the carbene–CH$_2$ distance is much shorter and around 1.95 Å. Although this in an unusual approach, the free energies obtained were in a quite good qualitative agreement with the experimental observations. The lowest free energy barrier was found for the alkene approaching the carbene through the *Re* face, with the phenyl group in a *trans* arrangement to the ester group.

Later on, Drudis-Solé et al. studied the cyclopropanation of Ph$_2$C=CH$_2$ using the catalyst shown in Fig. 13b with QM/MM ONIOM calculations [66]. In this case a different approximation was employed to locate the concerted transition state; the Gibbs free energy surface was explored with a small model system by using two reaction coordinates (x,y) directly related to the two forming C–C bonds (d_2 and d_3 in Fig. 13b). The x coordinate indicates the progress of the reaction, while y corresponds to the synchronicity of the transition state. Using this methodology, the concerted transition state was found to lie 5.7 kcal mol^{-1} above the reactants with coordinates $x = 3.6$ Å and $y = 0.0$, indicating a completely synchronous process.

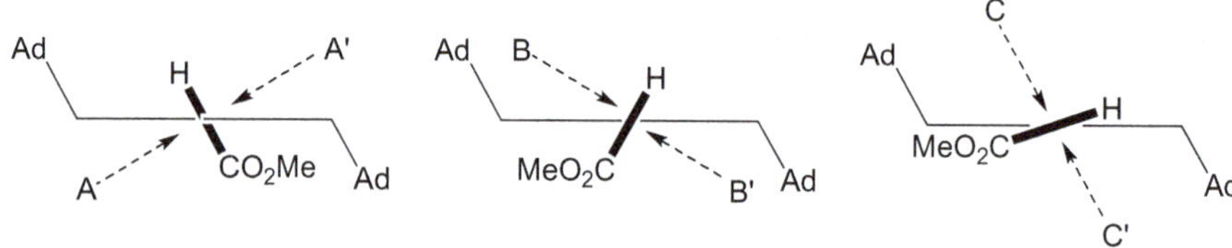

Fig. 13 QM/MM strategies to find the concerted transition state of cyclopropanation in the Gibbs energy surface of styrene (**a**) and Ph$_2$C=CH$_2$ (**b**) with copper complexes

Fig. 14 Possible relative positions of the carbene with respect to the Cu-Box plane in the cyclopropanation of Ph$_2$C=CH$_2$

The transformation of these coordinates back to C–C distances produces values of $d_2 = d_3 = 3.6$ Å.

Subsequently, they employed the same methodology for building the free energy surface of the full catalytic system using only x as the reaction coordinate and studying just synchronic pathways ($y = 0$). In this way 24 different reaction pathways were investigated considering three positions of the carbene (A, B, and C, Fig. 14) which could be attacked through the two faces and with different dihedral angles between the carbene and the oxy and methoxy groups of the ester. In all the cases, the R and S products can be obtained. Thus, the free energy profile was computed for all the pathways allowing the calculation of the enantiomeric excess for all cases. The use of a QM/MM approach was critical to enhance such a computationally demanding approach.

The most favorable pathway affords a 90% *ee* for the S product, in nice agreement with the 98% *ee* in S found experimentally. The enantioselectivity of this system seems to arise from the relative orientation of the carbene with respect to the bis-oxazoline ligand plane, the steric repulsions that appear between the adamantyl substituent and the ester group and between the phenyl groups of the alkene and both the adamantyl groups and the Box ligand.

García and coworkers employed a different strategy to study the enantioselectivity of cyclopropanation reactions where an enthalpic energy barrier cannot be located. This consists of forcing the system to have an enthalpy barrier by making it less reactive, i.e., decreasing the nucleophilic nature of the olefin [67]. They studied the cyclopropanation of styrene with the catalyst shown in Fig. 13a, with the same level of theory, but in this case they included the phenyl ring of the substrate in the MM part. This replacement increases the barrier of the reaction, and thus the concerted transition state could be located without the need of computing the Gibbs free energy surface. The computed enantioselectivity was similar as the one reported above.

7 Conclusions and Perspectives

QM/MM methods are already a widely established tool for the calculation of free energy profiles for complex systems in homogeneous catalysis. They are especially suited for mononuclear systems where the explicit introduction of bulky ligands in the calculation is necessary to reproduce the experimental behavior. The need to introduce the real bulky ligand explicitly in the calculation is more frequent in problems involving regioselectivity and enantioselectivity.

The proper characterization of the mechanistic origin of selectivity is a highly desirable target for computational chemistry. Selectivity is the key in many processes of practical interest, and the availability of a reliable predicting tool from computational chemistry is in high demand. QM/MM methods are a very valid tool for the study of reaction mechanisms in selective homogeneous catalysis.

References

1. Maseras F, Lledós A (eds) (2002) Computational modeling of homogeneous catalysis. Kluwer, Dordrecht
2. Thiel W (2014) Angew Chem Int Ed 53:8605
3. Bonney KJ, Schoenebeck F (2014) Chem Soc Rev 43:6609
4. Senn HM, Thiel W (2009) Angew Chem Int Ed 48:1198
5. Chung LW, Sameera WMC, Ramozzi R, Page AJ, Hatanaka M, Petrova GP, Harris TV, Li X, Ke ZF, Liu FY, Li HB, Ding LN, Morokuma K (2015) Chem Rev 115:5678
6. Maseras F (1999) Top Organomet Chem 4:165
7. Maseras F (2000) Chem Commun 1821
8. Maseras F (2001) In: Cundari TR (ed) Computational organometallic chemistry. Marcel Dekker, New York
9. Ujaque G, Maseras F (2004) Struct Bond 112:117
10. Bo C, Maseras F (2008) Dalton Trans 2911
11. Ananikov VP, Musaev DG, Morokuma K (2010) J Mol Catal A Chem 324:104
12. Sameera WMC, Maseras F (2012) WIREs Comput Mol Sci 2:375
13. Balcells D, Maseras F (2007) New J Chem 31:333

14. Brown JM, Deeth RJ (2009) Angew Chem Int Ed 48:4476
15. Krenske EH, Houk KN (2013) Acc Chem Res 46:979
16. Warshel A, Levitt M (1976) J Mol Biol 103:227
17. Field MJ, Bash PA, Karplus M (1990) J Comput Chem 11:700
18. Maseras F, Morokuma K (1995) J Comput Chem 16:1170
19. Svensson M, Humbel S, Froese RDJ, Matsubara T, Sieber S, Morokuma K (1996) J Phys Chem 100:19357
20. Dapprich S, Komaromi I, Byun KS, Morokuma K, Frisch MJ (1999) J Mol Struct (TheoChem) 461:1
21. Lin H, Truhlar DG (2007) Theor Chem Acc 117:185
22. Vreven T, Morokuma K (2003) Theor Chem Acc 109:125
23. Barea G, Lledós A, Maseras F, Jean Y (1998) Inorg Chem 37:3321
24. Assfeld X, Rivail JL (1996) Chem Phys Lett 263:100
25. Gao JL, Amara P, Alhambra C, Field MJ (1998) J Phys Chem A 102:4714
26. Walsh PJ, Kozlowski MC (2009) Fundamentals of asymmetric catalysis. University Science, Sausalito
27. Laidler KJ (1987) Chemical kinetics. Prentice Hall, New York
28. Kozuch S, Shaik S (2011) Acc Chem Res 44:101
29. Halpern J, Harrod JF, James BR (1961) J Am Chem Soc 83:753
30. Young JF, Osborn JA, Jardine FH, Wilkinson G (1965) Chem Commun 131
31. Nozaki H, Moriuti S, Takaya H, Noyori R (1966) Tetrahedron Lett 43:5239
32. Knowles WS, Sabacky MJ (1968) Chem Commun 1445
33. Dang TP, Kagan HBJ (1971) Chem Soc Chem Commun 481
34. Landis CR, Halpern J (1987) J Am Chem Soc 109:1746
35. Gridnev ID, Imamoto T (2004) Acc Chem Res 37:633
36. Daniel C, Koga N, Han J, Fu XY, Morokuma K (1988) J Am Chem Soc 110:3773
37. Landis CR, Hilfenhaus P, Feldgus S (1999) J Am Chem Soc 121:8741
38. Feldgus S, Landis CR (2000) J Am Chem Soc 122:12714
39. Landis CR, Feldgus S (2000) Angew Chem Int Ed 39:2863
40. Burk MJ, Feaster JE, Nugent WA, Harlow RL (1993) J Am Chem Soc 115:10125
41. Knowles WS (1983) Acc Chem Res 16:106
42. Mori S, Vreven T, Morokuma K (2006) Chem Asian J 1:391
43. Feldgus S, Landis CR (2001) Organometallics 22:2374
44. Donoghue PJ, Helquist P, Wiest O (2007) J Org Chem 72:839
45. Donald SMA, Vidal-Ferran A, Maseras F (2009) Can J Chem 87:1273
46. Fernández-Pérez H, Donald SMA, Munslow IJ, Benet-Buchholz J, Maseras F, Vidal-Ferran A (2010) Chem Eur J 16:6495
47. Shimizu H, Ishizaki T, Fujiwara T, Saito T (2004) Tetrahedron Asymmetry 15:2169
48. Donoghue PJ, Helquist P, Norrby PO, Wiest O (2009) J Am Chem Soc 131:410
49. Donoghue PJ, Helquist P, Norrby PO, Wiest O (2008) J Chem Theory Comput 4:1313
50. van Leeuwen PWNM, Claver C (eds) (2002) Rhodium catalyzed hydroformylation. Springer, Heidelberg
51. Franke R, Selent D, Boerner A (2012) Chem Rev 112:5675
52. Kégl T (2015) RSC Adv 5:4304
53. Schmidt S, Deglmann P, Hofmann P (2014) ACS Catal 4:3593
54. Jacobs I, de Bruin B, Reek JNH (2015) ChemCatChem 7:1708
55. Carbó JJ, Maseras F, Bo C, van Leeuwen PWNM (2001) J Am Chem Soc 123:7630
56. Zuidema E, Escorihuela L, Eichelsheim T, Carbó JJ, Bo C, Kamer PJK, van Leeuwen PWNM (2008) Chem Eur J 14:1843
57. Carbó JJ, Lledós A, Vogt D, Bo C (2006) Chem Eur J 12:1457
58. Aguado-Ullate S, Saureu S, Guasch L, Carbó JJ (2012) Chem Eur J 18:995
59. Aguado-Ullate S, Guasch L, Urbano-Cuadrado M, Bo C, Carbó JJ (2012) Catal Sci Technol 2:1694

60. Salomon RG, Kochi JK (1973) J Am Chem Soc 95:3300
61. Fraile JM, García JI, Martínez-Merino V, Mayoral JA, Salvatella L (2001) J Am Chem Soc 123:7616
62. Evans DA, Woerpel KA, Hinman MM, Faul MM (1991) J Am Chem Soc 113:726
63. Fraile JM, García JI, Mayoral JA, Tarnai T (1999) J Mol Catal A 144:85
64. Fraile JM, García JI, Gil MJ, Martínez-Merino V, Mayoral JA, Salvatella L (2004) Chem Eur J 10:758
65. Rasmussen T, Jensen JF, Østergaard N, Tanner D, Ziegler T, Norrby PO (2002) Chem Eur J 8:177
66. Drudis-Solé G, Maseras F, Lledós A, Vallribera A, Moreno-Mañas M (2008) Eur J Org Chem 5614
67. García JI, Jiménez-Osés G, Martínez-Merino V, Mayoral JA, Pires E, Villalba I (2007) Chem Eur J 13:4064

Struct Bond (2016) 167: 81–106
DOI: 10.1007/430_2015_183
© Springer International Publishing Switzerland 2015
Published online: 9 October 2015

Realistic Simulation of Organometallic Reactivity in Solution by Means of First-Principles Molecular Dynamics

Pietro Vidossich, Agustí Lledós, and Gregori Ujaque

Abstract The application of first-principles molecular dynamics simulations to the study of the reactivity of organometallic complexes is surveyed, with special emphasis on studies addressing catalytic processes. We focused on modeling studies in which the solvent, either water or nonaqueous, is explicitly represented. Where available, comparison is made with results obtained from static calculations based on reduced model systems (clusters). In doing so, we show how the mechanistic insight provided by modeling studies of reactions involving charge separation (e.g., proton release) or unsaturated species may qualitatively and quantitatively change when more extended model systems are considered. General aspects of the methodology are also presented.

Keywords AIMD • Explicit solvent • Homogeneous catalysis • Organometallic reactivity • QM/MM-MD

Contents

P. Vidossich (✉), A. Lledós (✉), and G. Ujaque (✉)
Departament de Química, Edifici C.n., Universitat Autònoma de Barcelona, 08193 Cerdanyola del Vallès, Catalonia, Spain
e-mail: vido@klingon.uab.es; agusti@klingon.uab.es; gregori.ujaque@uab.cat

Abbreviations

AIMD   Ab initio molecular dynamics
CPMD   Car–Parrinello molecular dynamics
DFT   Density functional theory
MD   Molecular dynamics
QM/MM   Quantum mechanics/molecular mechanics

1 Introduction

Organometallic chemistry is possibly one of the fields in which the application of computational quantum chemistry methods has been most successful [1–3]. Indeed, it is now common practice to complement synthetic or spectroscopic studies with the computational characterization of putative species present in reaction mixtures [4–6]. Detailed mechanistic studies, employing state-of-the-art electronic structure methods (notably, density functional theory, DFT), may also be performed to rationalize experimental outcomes [7, 8]. Many of these studies are based on reduced model systems, in which the reactive moieties are represented explicitly and environmental effects are included by means of mean field theories [9–11]. The approach has turned out to be successful in many instances and may be routinely performed on general-purpose hardware.

But, is a continuum representation of solvent always sufficient? May we neglect specific solute–solvent interactions such as H-bonds? And what about reactions in which the solvent plays the role of a reactant? Furthermore, do species in solution interact, e.g., a charged catalyst with its counterion, and how would this affect the catalyst's reactivity? To address these and similar questions, we first have to extend the model systems considered to include an explicit representation of solvent and of those molecular species which may have an impact on reactivity. A first step in this direction is the use of cluster–continuum models in which a reduced number of explicit solvent molecules are introduced in the model. This approach has been successfully applied in computational studies of the organometallic reactivity [12–14], yet it suffers from some limitations [15]. How many solvent molecules should be explicitly included? Is the first solvation sphere enough? In order to mimic bulk conditions, models have to include enough solvent molecules to fully solvate the solute. These models, which are built to reproduce experimental densities, are generally treated as periodically repeating units in order to remove the explicit/ continuum (or vacuum) boundary.

Dealing with large model systems however poses a challenge: how to explore its configurational space? Systematically investigating how the potential energy varies as a function of few nuclear coordinates, as is currently done in computational chemistry studies based on reduced model systems, is not viable because of the large number of particles involved. Statistical mechanics techniques, such as molecular dynamics or Monte Carlo, have to be used to sample the many configurations accessible by the system. The present review is an account of the use of molecular dynamics simulations to investigate organometallic reactions (see, e.g., [16] for an overview of the capabilities of first-principles MD simulations applied to chemical reactivity). A pioneering application of ab initio molecular dynamics (AIMD) simulations to the study of organometallic reactivity was published in the mid-1990s by Ziegler's group [17]. Because of the reasons outlined above, we focused on studies in which solvent was explicitly represented (see, e.g., [18] for an account of studies of organometallic systems in isolation). The material presented in this review is intended to highlight the relevance of including explicitly the solvent and what may be gained in doing so. We present selected mechanistic studies of several classes of reactions. It will become apparent to the reader that most studies investigated reactions performed in water. This certainly reflects the peculiar properties of water, including acid–base and coordinative properties. However, explicit modeling of other protic solvents (e.g., alcohols) and even apparently inert solvents such as toluene may be required to properly describe certain processes. We start with an accessible introduction of the methodology employed.

2 Methodological Aspects

Molecular dynamics (MD) is a simulation technique used to follow the time evolution of a molecular system [19, 20]. Within a classical representation of nuclei, the technique consists of solving numerically Newton's equations of motion [Eq. (1)] for all atoms in the system:

$$m_I \frac{d^2 \mathbf{R}_I}{dt^2} = \mathbf{F}_I(\{\mathbf{R}_I\}) = -\nabla_I E(\{\mathbf{R}_I\}) \qquad I = 1 \cdots N \tag{1}$$

where m_I is the mass of atom I, $\mathbf{R}_I$ its position vector, $\mathbf{F}_I$ the force acting on it, and E the interaction potential function which depends on the positions of all the nuclei in the system. Several schemes have been proposed for a stable solution of Eq. (1), a popular one being Verlet's algorithm [21], which predicts the position at time $\mathbf{R}_I(t + \Delta t)$ based on the position at time t and $t - \Delta t$ and the acceleration at time t:

$$\mathbf{R}_I(t + \Delta t) = 2\mathbf{R}(t) - \mathbf{R}_I(t - \Delta t) + \frac{\mathbf{F}_I(t)}{m_I} \Delta t^2 \tag{2}$$

The time step Δt used to solve Eq. (2) is a major parameter determining the stability of the simulation. For a proper propagation, Δt has to be chosen to be compatible with the highest vibrational modes of the system. When these involve bond stretching of hydrogen atoms, a time step up to 0.5 fs is appropriate. With a proper choice of Δt, solutions of Eq. (2) conserve the total energy of the system [19].

Simulations generally involve a fixed number of particles (N). To model bulk systems with a finite number of particles, periodic boundary conditions are applied. The volume (V) of the periodically repeating simulation box may be kept fixed or varied. Extensions to the scheme in Eq. (2) allow coupling the system to a thermostat and/or barostat to control the temperature T and pressure P of the system [22–24]. Hence, depending on the chosen conditions, the distribution sampled during the MD simulation is representative of either the microcanonical (NVE), canonical (NVT), or isothermal–isobaric (NPT) statistical mechanics ensembles [19].

Force evaluation is required for propagating nuclear positions [Eq. (2)]. This operation is repeated at each time step and has a major impact on the cost of the simulation. Forces are obtained as the gradients of some potential energy function E [Eq. (1)]. For large molecular systems, the so-called on-the-fly approach consists in calculating forces for each configuration $\boldsymbol{R}_I$ visited by the system during the simulation. When forces are computed from first principles, the procedure is known as ab initio molecular dynamics (AIMD) [25]. Within the Born–Oppenheimer approximation, forces are computed after optimizing the wave function at each step during the dynamics (Born–Oppenheimer AIMD). To avoid this costly evaluation, Car and Parrinello developed an efficient and accurate scheme, according to which the orbitals are treated as classical particles and are propagated simultaneously with the ions [26].

Among the electronic structure methods to be used in AIMD simulations, density functional theory (DFT) [27], because of accuracy and favorable scaling with the number of atoms, is the recommended method to model organometallic systems. Extensive research has been dedicated to test and improve the performance of DFT when dealing with transition metal systems, and dedicated reviews cover recent developments [28–30].

According to the Kohn–Sham formulation of DFT [31], the energy of the system is given by Eq. 3

$$E = \min_{\{\varphi_i\}} E_{\mathrm{KS}}\left[\{\varphi_i(\boldsymbol{r})\}; \{\boldsymbol{R}_I\}\right] + V_{\mathrm{NN}}(\{\boldsymbol{R}_I\}) \tag{3}$$

where the E_{KS} functional contains kinetic, nuclear-electron, and electron–electron interaction energy terms, $\varphi_i(\boldsymbol{r})$ are the molecular orbitals, and V_{NN} refers to the nuclear–nuclear interaction energy. The electron–electron term includes classical (Coulomb) and nonclassical interactions [32, 33]. The latter is accounted for by the exchange–correlation functional, E_{xc}, of which only approximate forms are known. Because of its approximate nature, E_{xc} is the main source of error in DFT calculations and thus of AIMD simulations. Some of the well-known deficiencies of E_{xc}

include the neglect of dispersion interactions, which are responsible for the underestimation of interaction energies between apolar fragments, and the so-called self-interaction error, which is responsible for the unphysical electron delocalization experienced in some open-shell systems [34]. Newly parameterized E_{xc} functionals [35], or the inclusion of empirical correction terms [36, 37], are capable of better describing van der Waals complexes. The inclusion of a fraction of exact (Hartree–Fock) exchange in the functional decreases the self-exchange interaction, improving the description of reaction barriers and open-shell systems [30]. However, in problematic cases, specific corrections have to be introduced as a remedy to the self-interaction error [38, 39]. DFT is a ground state theory, but extensions to treat excited states are available. Among these, time-dependent DFT (TD-DFT) found widespread application for the study of electronic transitions [40, 41].

Despite efforts to develop more efficient schemes (most notably, linear scaling methods) [42], the treatment of large systems by QM methods is computationally intensive (see [43, 44] for some notable examples of AIMD applications to systems of up to 1,000 atoms). To maintain the computational cost tractable, the multiscale quantum mechanical/molecular mechanical (QM/MM) approach may be used [45–47]. The approach, well established in the field of computational biochemistry [48, 49], consists in accurately modeling the chemistry at the reactive center (the organometallic frame, described at the QM level), while accounting for environmental (solvent) effects at the MM level. Since the original proposal, researchers have proposed different schemes to couple the QM and the MM subsystems (see, e.g., [48] for a comprehensive account). This approach has been extensively applied to modeling organometallic reactivity of systems with bulky substituents by means of traditional static methods, usually describing the reactive part at the QM level and the ligand and substrate substituents at the MM level [50–52]. Since then, developments to combine QM/MM potentials with ab initio molecular dynamics (QM/MM-MD) for the study of organometallic systems were described [53–57]. In the present context, the QM region may include solvent molecules if these are supposed to participate in the reaction. When these molecules are not strongly bound to the reactive moiety, diffusion may take place. To prevent exchange of QM and MM solvent molecules, either geometrical restraints have to be applied to the system [58] or sophisticated adaptive QM/MM partitioning schemes have to be adopted [59, 60].

Once the system is setup, the AIMD simulation is launched and extended for some amount of simulated time. As pointed out above, the system will visit states with a probability determined by the simulation conditions, e.g., Boltzmann statistics for a simulation performed at constant temperature. Accordingly, high-energy states, such as transition states, will be observed with low probability compared to the bottom of the free energy surface, which will be sampled more extensively. Because of the cost of AIMD simulations, the total accessible simulated time (of the order of hundreds of ps) is limited, and thus transitions (conformational transitions or chemical rearrangements) will likely not be observed during the simulation. A rough estimate of the simulated time required to observe a transition between two states separated by an energy barrier $\Delta G^{\ddagger}$ is $1/k^{TST}$, where k^{TST} is the transition state theory estimate of the rate constant [61]. To overcome this limitation, enhanced sampling methods have been proposed [62]. Biased methods are particularly suitable to study chemical reactions, with thermodynamic integration [63] and metadynamics [64–66] being the ones that found widespread application in the field of AIMD. In both methods, one or few variables (hereafter, s), defined in terms of atomic positions, are used to drive the reaction. This set of variables is chosen as a representation of the reaction coordinate and thus is required to distinguish reactants from product (and intermediate) states. The bias acts to improve the sampling of these variables. In thermodynamic integration, a constraint is applied to maintain the reactive variable(s) at fixed values. The average force acting on the constraint in each constrained ensemble (the method is also known as constrained dynamics) [67, 68] is used to reconstruct the free energy of the process according to Eq. (4):

$$\Delta F(s) = -\int_0^s ds' \left\langle \frac{\partial H}{\partial s} \right\rangle_{s'} \tag{4}$$

In metadynamics, a history-dependent repulsive potential pushes the system away from stable states [69]. The potential is built along the dynamics as a sum of Gaussian functions which are added to the physical potential at intervals t':

$$V(s, t) = \sum_{t'} h exp \left[\frac{-(s(t) - s(t'))^2}{2w^2} \right] \tag{5}$$

where w is the width and h the height of the Gaussian functions. In the limit of $t \to \infty$, the (negative of) $V(s,t)$ approximates the underlying free energy. Guidelines on how to choose w and h may be found in [70, 71]. The convergence properties of $V(s,t)$ were studied in [72].

Because of the local nature of most chemical reactions, distances between reactive atoms are intuitive and appropriate choices for the s variables. It should be taken into account, however, that the cost of estimating free energy changes increases with the number of variables used to drive the reaction. Combination of variables may be an effective way to relieve this issue. For instance, in an S_N2 reaction, the difference of the distances of attacking and leaving group from the transferring group allows the use of a single variable to describe the process. A further issue to be taken into account is when the reactive species include chemically equivalent species as it occurs, for instance, in reactions in which water acts as the nucleophile. When bulk conditions are applied, any water molecule may attack the substrate. An appropriate s variable to take this into account is the coordination number (CN) [73]:

$$CN = \sum_i \frac{1 - (r_i/r_0)^p}{1 - (r_i/r_0)^q} \tag{6}$$

where r_i is the distance of the water oxygen atom i from the substrate atom and parameters p, q, and r_0 define the decay of the rational function.

It is important to note that the trajectories generated from biased simulations are not reactive trajectories. However, the biased simulation is interpreted in the sense that a reactive trajectory would pass through the same states visited in the biased simulation. Genuine reactive trajectories may be obtained launching simulations from high-energy configurations observed in the biased simulation [74].

Many popular quantum chemistry codes now include the possibility of performing AIMD simulations. Because of the need to propagate the equations of motion several thousands of times, highly efficient codes are required to perform AIMD simulations of complex molecular systems. The authors are more familiar with the CPMD (www.cpmd.org) and CP2K (www.cp2k.org) program packages, which were designed for atomistic simulations of large systems, and the applications presented in this review are mostly based on these codes. Both codes are particularly suited to high-performance computing resources but display good performances on general-purpose clusters provided with tightly coupled interprocess communications [75, 76]. Other codes implementing AIMD include VASP (www.vasp.at), GAUSSIAN (www.gaussian.com), deMon (www.demon-software.com), and Quantum ESPRESSO (www.quantum-espresso.org).

3 Applications

In this section, we illustrate applications of first-principles molecular dynamics simulations to the study of the reactivity of organometallic complexes in explicit solvent. We mostly focus on catalytic processes, although not always complete catalytic cycles have been addressed. We found convenient to distinguish reactivity in water (covered in Sect. 3.1) from nonaqueous solvents (covered in Sect. 3.2). This choice was made because of the prominent role of studies of aqueous systems in the field of first-principles simulations [77]. However, homogeneous catalytic processes are mostly developed in organic solvents, and the range of reactions experimentally studied is actually more diverse when we come to this class of solvents.

3.1 Reactivity in Water

Oxidation reactions in water are among the most studied processes by means of explicit solvent AIMD. In the Wacker, water splitting, and Shilov processes, proton release to the bulk solution is observed, a charge separation process which can be properly described in explicit solvent models. In the Fenton system, concerted radical propagation and proton transfer through the aqueous medium facilitate the formation of the high-valent iron catalysts. Cisplatin hydrolysis, a ligand exchange reaction, has been included as a prototype study addressing the coordinative properties of water, which are of high relevance for medicine-oriented design of organometallic systems. The hydrogenation reaction promoted by ruthenium catalysts constitutes a clear example in which the presence of explicit solvent molecules stabilized charged intermediates. Further studies in which the effect of pH on the reactivity of transition metal systems is studied conclude this section.

3.1.1 The Fenton Chemistry

Fenton chemistry refers to an oxidation process that uses H_2O_2 as oxidant and a transition metal catalyst, generally Fe, to oxidize organic substrates. This process is commonly used for wastewater treatment. Two reaction mechanisms were proposed for the Fenton reagent. One assumes the production of free hydroxyl radicals by the metal-catalyzed decomposition of peroxide [78], whereas the other involves the formation of an iron–oxo complex, such as $[Fe^{IV}O(OH)(H_2O)_4]^{2+}$ (named hereafter $[Fe^{IV}O]^{2+}$), as a highly reactive oxidative intermediate [79].

Baerends and coworkers have largely contributed to the understanding of the Fenton process at the atomistic level. In a first contribution using explicit water AIMD simulations, they obtained important conclusions not affordable by means of static calculations [80]. They considered a cubic cell of ~10 Å edge including Fe^{2+}, H_2O_2, and 31 water molecules periodically repeating in space. The explicit representation of solvent allowed proper solvation of reacting species (Fe^{2+} and reaction intermediates) and, importantly, propagation of OH^-, H^+, and OH· through the solution.

Simulations established that the peroxide dissociates into two OH groups. One OH radical coordinates to the iron center. The other attacks a water molecule from the bulk rather than an iron-coordinated water molecule as suggested by calculations performed in gas phase. In a fast chain reaction, the OH· is passed on via two solvent water molecules, and the reaction is completed by abstraction of a hydrogen atom from an aqua ligand of the iron complex in the neighboring cell. The overall process was described as a concerted reaction rather than a chain reaction since the abstraction of a hydrogen atom from a coordinated aqua ligand occurred simultaneously with the breaking of the O–O bond. Formally, the overall process can be represented by

$$\left[Fe^{II}(H_2O)_5(H_2O_2)\right]^{2+} \rightarrow \left[Fe^{IV}(OH)_2(H_2O)_4\right]^{2+} + H_2O \qquad (7)$$

The Fe(IV) dihydroxide complex further evolved in solution to the formation of the $[Fe^{IV}O]^{2+}$ complex with the transfer of a proton to the bulk solution:

$$\left[Fe^{IV}(OH)_2(H_2O)_4\right]^{2+} + H_2O \rightarrow \left[Fe^{IV}O(OH)(H_2O)_4\right]^{+} + H_3O^+ \qquad (8)$$

Once again, the presence of explicit solvent was crucial to represent this pathway because it enabled an H^+ to be stabilized by the solvent molecules. In a subsequent study, the same group analyzed more deeply the formation of the $[Fe^{IV}O]^{2+}$

intermediate, corroborating that the process takes place in two steps and showing that from an energetic point of view, formation of the ferryl ion is more favorable than formation of free OH· radicals [81]. Recently, Yamamoto et al. [82] reinvestigated the mechanism of peroxide activation in $[Fe^{II}(H_2O_2)(H_2O)_5]^{2+}$ and confirmed the findings from the Baerends group.

The reactivity of the $[Fe^{IV}O]^{2+}$ species toward organic substrates was then investigated [83]. Using a model of similar size as the one used to model $[Fe^{IV}O]^{2+}$ formation, Baerends and coworkers performed constrained molecular dynamics simulations to estimate the energetics of two reaction mechanisms. One involved methane coordination, whereas the other, known as the oxygen-rebound mechanism (which is related to the proposed mechanism of methane oxidation by monooxygenase and cytochrome P450 enzymes) [84], implies an H abstraction to generate a CH_3· radical species, which then reacts with the iron complex to give the CH_3OH product. The two pathways were first evaluated by means of static methods. The mechanism based on methane coordination was found to be high in energy and was thus discarded by the authors. The rebound mechanism was modeled including the solvent explicitly. The free energy barrier for the H abstraction step was calculated to be 21.8 kcal mol^{-1}, much higher than in vacuum (3.4 kcal mol^{-1}). The spin density was mapped along the pathway to monitor to electronic state of the iron catalyst. Antiferromagnetic coupling between the reacting species was reported, with the iron complex in the high-spin $S = 5/2$ state.

The mechanism for methanol oxidation to formaldehyde by the $[Fe^{IV}O]^{2+}$ species was also investigated [85], as a prototype reaction for alcohol oxidation in Fenton chemistry. Two pathways were considered, with the $[Fe^{IV}O]^{2+}$ species attacking either the C–H bond or the O–H bond. Considerable differences in the energetics computed from gas phase and solution models were reported. The C–H bond pathway in the gas phase turned out to have a low barrier of 0.5 kcal mol^{-1}. In solution, however, the free energy barrier was estimated to have an upper bound of 12 kcal mol^{-1}. The subsequent step involving the O–H bond breaking proceeded spontaneously. The alternative pathway that involves coordination of MeOH and hydrogen transfer from the coordinated OH group could not be discarded because the energy barrier was found to be 10.5 ± 2.4 kcal mol^{-1}. The reactivity was proposed to be associated with the shape and energy of the LUMO of the $[Fe^{IV}O]^{2+}$ species (the Fe–O σ^* orbital), suggesting that this species is a strong electron acceptor rather than an H-bond acceptor, thus facilitating the C–H bond activation of the organic compound. These ideas were recently further developed for the mechanism of C–H activation by $[EDTAH_n \cdot FeO]^{(n-2)}$ species [86].

3.1.2 The Wacker Oxidation

The Wacker process is a paradigmatic example of homogeneous catalysis applied in the chemical industry [87]. The process, discovered in the 1950s, produces acetaldehyde from ethene and oxygen using a Pd catalyst. The reaction is proposed to proceed via the addition of Pd and a hydroxide group (originated from the

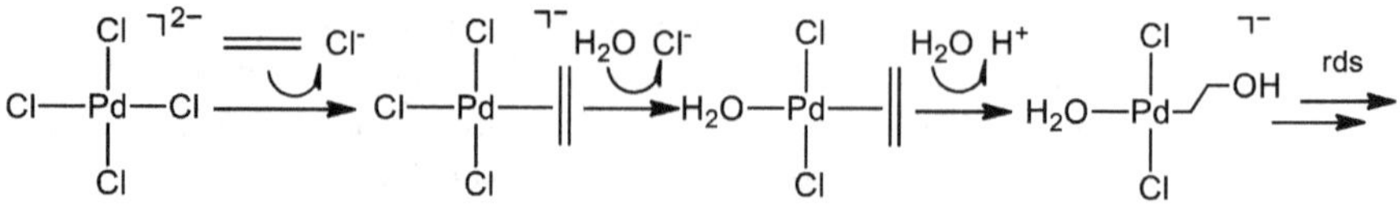

                                                                P. Vidossich et al.

Fig. 1 Proposed sequence of steps for hydroxypalladation of ethene in the Wacker process

nucleophilic addition of a water molecule) across the C=C bond, a process named "hydroxypalladation" (Fig. 1) [88]. The mechanism of this step, crucial for the reaction, remained controversial over many years until recently [89].

A number of studies investigated the reaction mechanism by means of static methods [88, 89]. Many of these studies included a cluster of water molecules for a better modeling of the system, in particular of the proton transfer during the nucleophilic attack of a water molecule on the coordinated ethene. However, the use of an extended cluster of water molecules complicates the exploration of the potential energy surface due to the existence of several manifolds with similar energy. To overcome this limitation, free energy methods that account for finite temperature effects are required.

An important issue in the Wacker mechanism concerns the nature of the catalytically active Pd^{II} species. Ligand exchange at $[PdCl_4]^{2-}$ was studied by Kovács et al. [90] and Nair and coworkers [91, 92] by means of AIMD simulations in explicit water. Similar protocols were used, applying periodic boundary conditions to a cell of ~10 Å edge including the reactants and about 35 water molecules. The reactive events were explored by means of metadynamics.

The experimental rate law indicates that at low $[Cl^-]$, two Cl^- ligands are substituted by H_2O and C_2H_4. The *trans*-effect will determine the configuration of the $[PdCl_2(C_2H_4)(H_2O)]$ species. Calculations agree that the most likely pathway starts with the substitution of Cl^- by C_2H_4, followed by the exchange of a second Cl^- ligand for H_2O, generating *trans*-$[PdCl_2(C_2H_4)(H_2O)]$ (see Fig. 1). Remarkably, formation of the *trans* intermediate is of a considerable relevance for the reaction mechanism, because it implies the anti-nucleophilic addition by a water from the bulk (outer sphere mechanism), rather than the syn addition by a Pd-coordinated aqua ligand (inner sphere mechanism). Thus, explicit solvent simulations point to a more favored outer sphere mechanism. Importantly, the nucleophilic addition of water, leading to the formation of the C–O bond, takes place simultaneously to the proton release from the attacking water to the bulk solution.

The consequence of these observations is that the rate-determining step (rds) of the process should follow the hydroxypalladation to be in agreement with the experimental rate law. Bäckvall et al. [93] proposed a ligand dissociation step after the hydroxypalladation as the rds, although a β–hydrogen transfer, to form a Pd-hydride intermediate, has also been proposed [94] (although not supported by KIEs) [95]. The complete catalytic cycle was evaluated by means of metadynamics simulations for both forward and backward reactions (Fig. 2). The computed free energy profile showed that the rds involves *trans*-to-*cis* isomerization of Cl^- ligands and β–hydrogen elimination. The free energy barrier was associated with

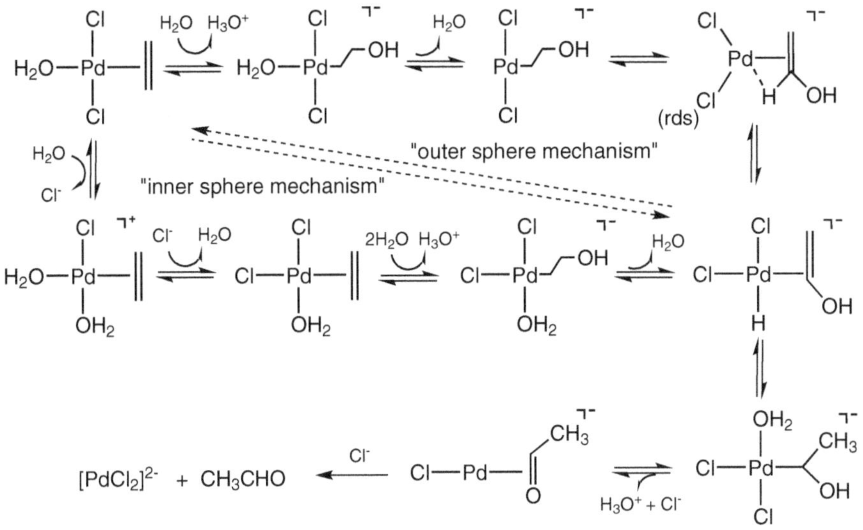

Fig. 2 Calculated reaction mechanisms for the Wacker process by AIMD simulations

the isomerization, because the formation of the Pd-hydride intermediate occurred immediately after as a barrierless process.

Another interesting part of the process concerns the formation of the aldehyde. This step involves the deprotonation of the OH group of the hydroxyethyl ligand. Several pathways were proposed in the literature for this step, including β–hydrogen elimination [96] and deprotonation by a Cl⁻ ligand [97] or by a water molecule from solution [91]. The three possibilities were accounted for in a metadynamics simulation, showing that the deprotonation occurs to a solvent water molecule (Fig. 2).

Interestingly, the free energy barriers of all the individual steps characterized during the simulation were used to build a kinetic model for the process. The experimental rate law could be reproduced using the values derived from first-principles molecular dynamics calculations [98].

3.1.3 Hydrolysis of Cisplatin

Cisplatin, cis-[Pt(NH$_3$)$_2$Cl$_2$], is a compound of considerable relevance in medicine due to its application as an anticancer drug. The activity of cisplatin is due to coordination properties of the Pt(NH$_3$)$_2$ moiety, which is capable of binding DNA. Ligand exchange at the Pt center is thus an important aspect of drug activation. Cisplatin in aqueous environments exchanges Cl⁻ for aqua ligands (Fig. 3), and this reactivity was among the first applications of AIMD to investigate transition metal complexes.

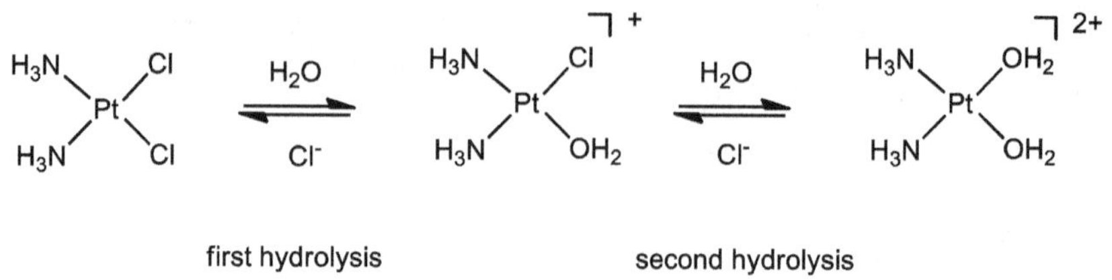

Fig. 3 Evaluated steps in the hydrolysis of cisplatin by CPMD calculations

Early studies using the Car–Parrinello approach were performed by Carloni et al. [99] The structure and dynamics of cisplatin in vacuum and water solution, as well as of the monoaqua complex, cis-$[Pt(NH_3)_2(H_2O)Cl]^+$, were analyzed. A periodically repeated cubic cell of 10.5 Å edge, including cisplatin and 35 water molecules, was considered. Importantly, the energies of the cisplatin and the monoaqua complexes were found to be the same within 1 kcal mol^{-1}. Hence, the presence of explicit water molecules, as a source of hydrogen bonds stabilizing the chloride ion, made the ligand substitution energetically accessible. In the gas phase, formation of the monoaqua complex would be highly endothermic because of the formation of separated charged species.

The free energy barrier for the chloride-by-aqua ligand substitution process was calculated via constrained molecular dynamics simulations at room temperature along an associative pathway. The estimated barrier of 21 kcal·mol^{-1} compared remarkably well with those obtained in several independent experiments, ranging from 23 to 26 kcal mol^{-1} ([100] and cites there in). The structure and dynamics of an adduct cisplatin–DNA was also investigated in water solution [99].

These studies were later extended by Lau and Ensing [101] by analyzing the two successive chloride-by-aqua ligand substitution process to form the cis-$[Pt(NH_3)_2(H_2O)_2]^{2+}$ complex from cisplatin. The ligand substitution processes were promoted by means of metadynamics, and umbrella sampling simulations were used to refine the free energy profiles. No assumptions on the nature of the mechanism (associative versus dissociative) were made in the choice of the variables used to drive the reaction. The free energy barriers obtained for the first and second chloride-by-aqua ligand substitutions were 23.3 and 18.5 kcal mol^{-1}, respectively; the relative free energies of the mono- and di-aqua complexes compared to cisplatin were 2.1 and −2.9 kcal mol^{-1}, respectively. The free energy estimates of both hydrolysis reactions of cisplatin were in very good agreement with experiments. Hydrolysis took place in a concerted manner, and tri- or penta-coordinated intermediates did not correspond to energy minima on the free energy profile. Interestingly, the dissociated Cl$^-$ ion remained close (~4.5 Å) to the Pt complex, forming an ion-pair throughout the simulation. The hydration shell of cisplatin as well as of the mono- and di-aqua complexes have been studied by Sánchez-Marcos' group by means of classical MD simulations, but this study is out of the scope of this review [102].

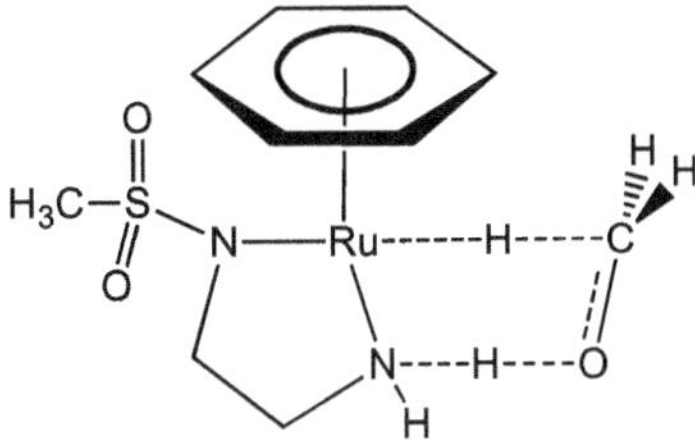

Fig. 4 The hydrogen transfer step studied by Pavlova and Meijer [101]

3.1.4 Ru-Catalyzed Hydrogen Transfer Process

The asymmetric hydrogenation of C=O and C=N bonds to generate stereogenic carbon centers is a transformation of great relevance for synthetic chemistry. The pioneering work of Noyori, who developed chiral Ru-based catalysts, was crucial in this field [103].

The key step of the process corresponds to the H_2 transfer from the Ru-based catalyst to the carbonyl group. Pavlova and Meijer studied this step by means of Car–Parrinello MD simulations using formaldehyde as the substrate, resulting in the formation of ethanol (Fig. 4) [104].

The authors used constrained dynamics to estimate the free energy barrier of the hydrogen transfer step. The reaction was studied in gas phase and water solvent. The calculated barrier for the hydrogenation of formaldehyde in water turned to be $3.4\,kcal\,mol^{-1}$, slightly lower than that obtained in gas phase, $5.3\,kcal\,mol^{-1}$. It was shown that when the hydride is transferred from the catalyst to the formaldehyde, the proton remains on the catalyst. The resulting methoxide intermediate was stabilized by multiple hydrogen bonds from surrounding water molecules. The proton transfer to the substrate was observed to take place from water molecules. Thus, explicit solvent simulations suggested a stepwise process, in agreement with kinetic isotope effect studies. Remarkably, this result is considerably different from that obtained in the gas phase, where the hydride and proton transfers were found to be concerted; recent studies combining cluster model with continuum methods (in alcohol as solvent) show a stepwise process (see Sect. 3.2.1).

The structural, energetic, and dynamical properties of bifunctional arene ruthenium catalysts have been investigated by means of classical and ab initio molecular dynamics simulations, although their reactivity was not addressed [105]. The hydrogen transfer reaction involving $NaBH_4$ hydrolysis has also been studied by means of explicit solvent AIMD [106].

3.1.5 Water Splitting Processes

The splitting of water into O_2 and H_2 is a major objective for the storage of solar energy into chemical fuels [107]. Two half-reactions are involved, namely, water oxidation, releasing O_2, H^+, and e^-, and reduction of protons to H_2. Water oxidation is a challenging process because of its multi-electron and multi-proton nature.

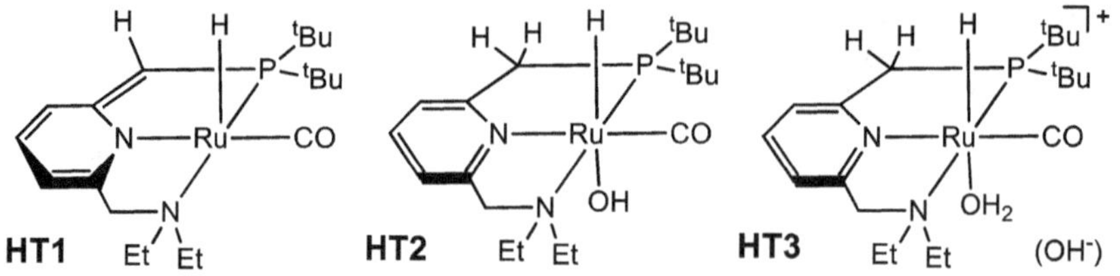

Fig. 5 Schematic representation of the Ru-pincer catalyst and the two pincer-aromatized species proposed to form in solution on the basis of static and dynamic calculations

The discovery of Ru(II)-pincer complex as catalyst for this process demonstrated that a single metal center can promote the reaction [108]. Fabris and coworkers studied by means of AIMD calculations the reaction mechanism, illustrating the importance of describing the solvent explicitly [109]. Calculations were performed in a periodic cell (~13.5 Å edge) containing the catalyst and 73 water molecules. Reactive steps were promoted by means of metadynamics. Experimental measurements suggested that water addition to the Ru-pincer complex (HT1 in Fig. 5) yields HT2, featuring the aromatization of the pincer ligand by addition of a H^+ to the ligand and a hydroxo ligand to the metal. AIMD simulations based on the HT2 complex in explicit water showed that the hydroxo ligand spontaneously converted into an aqua ligand leaving a hydroxide in solution (HT3 in Fig. 5). Based on these results, the authors proposed an alternative route for the aromatization of the pincer ligand. Accordingly, the Ru-pincer complex coordinates a water ligand and takes an additional H^+ from a water molecule of the bulk solution. Based on simulations, the estimated activation free energy was found to be 8.8 kcal mol^{-1}.

H_2 formation was also investigated. H_2 forms from the addition of a proton to the Ru–H hydride. Two possible sources for the proton were considered: one in which the proton was coming from the pincer ligand (requiring a solvent water molecule to act as a proton shuttle) and the other in which the proton was coming from a solvent water molecule. The former route requires dearomatization–aromatization steps within the complete catalytic cycle. The latter route, which does not require the dearomatization–aromatization of the ligand, turned to be lower in energy, with a barrier of 20 kcal mol^{-1} (proton transfer from the ligand featured a barrier of 31 kcal mol^{-1}). These results allowed the authors to propose a new reaction mechanism where the Ru(II)-pincer always retains its aromatic character throughout the process.

The activity of other Ru-based catalysts for water oxidation has been investigated. Buda and coworkers addressed the reactivity of $[(Ar)Ru(X)(bpy)]^+$ (with X^- replaced by a water molecule) by means of AIMD calculations. The model system contained the catalyst and 73 water molecules in a periodic box of $16 \times 15.5 \times 15$ Å^3 [110]. The authors assumed a catalytic cycle in which the X^- ligand is initially replaced by a water molecule and the formed intermediate then undergoes two one-electron oxidation steps in which two protons are released leading to a ruthenium(IV)–oxo complex, $[(Ar)Ru(O)(bpy)]^{2+}$, with a strong radical character on the oxyl ligand. This species is supposed to be the active species in the formation of the O–O bond with the nucleophilic attack of a water molecule onto the oxo ligand. This reaction step was the focus of the analysis by the authors. Two different reaction mechanisms were investigated for the formation of the O–O bond: one assuming that the incoming water molecule first binds to the Ru metal center and the other considering that the water molecule directly attacks to the oxyl ligand. Simulations show that water coordination to the oxo species was not viable because it leads to the breaking of the metal-aromatic ligand bond. The alternative pathway gives rise to the formation of the O–O bond generating a Ru-hydroperoxo complex; during this process, a proton is released to the water solution. The process was interpreted as a proton-coupled electron transfer on the basis of a charge distribution analysis. The aromatic ligand was suggested to play an important role as an electron charge buffer facilitating electron density adjustments during the catalytic cycle.

Buda's group also analyzed water oxidation as catalyzed by a di-Mn complex, $[(bis(imino)pyridine)(H_2O)Mn^{IV}(\mu\text{-}O)_2Mn^{V}(O)(bis(imino)pyridine)]^{3+}$ [111]. QM/MM-MD simulations were performed in a cubic box of 33 Å edge, containing the metal complex and around 1,000 water molecules. The di-Mn complex and three or four water molecules were included in the QM region, whereas the rest of the system was described at MM level. Full QM AIMD simulations were also performed in a periodic cell of $20.5 \times 23.8 \times 18.1$ Å^3, including 72 water molecules. Both setups consistently showed that the complex undergoes a structural rearrangement involving the release of the coordinated water molecule and the breakage of one of the oxo-bridges, thus generating two penta-coordinated Mn complexes bridged by an oxo ligand (Fig. 6a). Analysis of a subsequent reaction step for the O–O bond formation allowed the authors to conclude that both oxyl ligands at the Mn centers are involved in the process: one of them is attacked by a water molecule leading to the formation of a hydroperoxo ligand, whereas the other accepts a proton from the attacking water, thereby keeping the total charge of the complex unchanged (Fig. 6b).

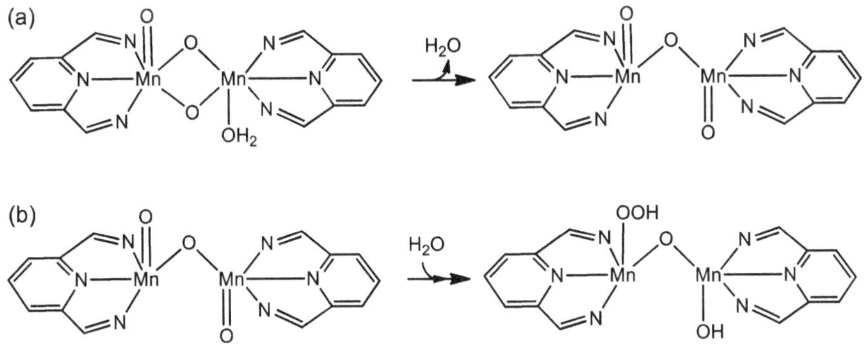

Fig. 6 (a) Rearrangement of the di-Mn structure in the solvent environment. (b) Reaction step for the O–O bond formation between the di-Mn complex and a water molecule [111]

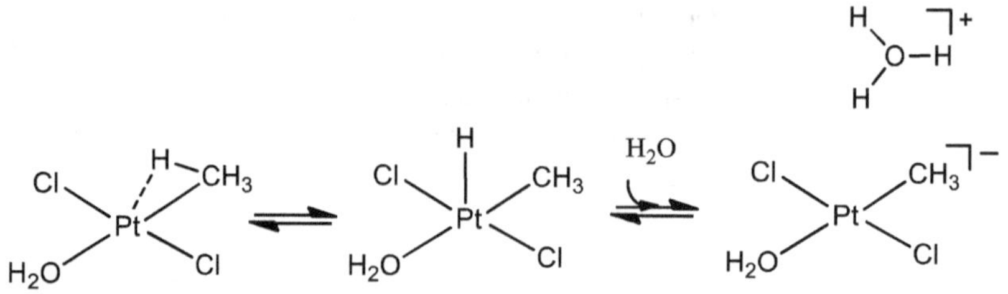

Fig. 7 Methane C–H bond activation by Pt^{II}

3.1.6 The Shilov Reaction

Alkane functionalization is a highly desirable chemical transformation. The Shilov system [112], based on Pt^{II} and Pt^{IV} complexes, constitutes an early proposal for hydrocarbon oxidation in aqueous solution (Fig. 7). The rate-limiting step of the catalytic cycle, namely, C–H bond activation, was investigated by Vidossich et al. by means of Car–Parrinello molecular dynamics simulations in explicit water [113]. The authors considered the PtCl$_2$(H$_2$O)(CH$_4$) species, in both *cis* and *trans* configurations, solvated with 32 water molecules in a cubic box of ~10 Å edge treated under periodic boundary conditions.

The *trans*-PtCl$_2$(H$_2$O)(CH$_4$) species turned not to be stable and spontaneously evolved to the methyl complex in the early stages of the simulation. Initially, the oxidative addition of the C–H bond to the Pt center resulted in a square pyramidal complex. Subsequently, the apical proton was released to a water molecule of the hydration sphere, resulting in the formation of a square planar methyl complex. On the contrary, the *cis* complex was maintained during the timespan of the simulation. The breakage of the C–H bond was thus promoted via metadynamics and required 6.4 kcal mol^{-1} free energy of activation. Similarly to the *trans* isomer, the proton moved to the Pt center to form a square pyramidal Pt complex, which spontaneously evolved to the square planar methyl complex by the release of the proton to the bulk solution. The activation barriers of the *cis* and *trans* σ-complexes were in line with previous work based on static calculations, and their difference was rationalized in terms of the effect of the ligand *trans* to methane. However, the high acidity of the five-coordinated hydride intermediate resulting from C–H activation could be revealed only by the use of explicit solvent models. Remarkably, the simulations showed that release of a proton to the hydration sphere was faster than water coordination to stabilize the formally Pt^{IV} center.

3.1.7 Other Processes

This subsection covers other important processes related to the reactivity or behavior of transition metal complexes in water solvent.

Reductive dehalogenation of chloroalkenes is an important process within the context of detoxification of halogenated pollutants. Bühl and Golubnychiy [114] analyzed the evolution of a relevant intermediate generated in a Co-catalyzed dehalogenation process. Coordination of $Cl_2C=CCl_2$ to $[Co(glyoximato)_2]^-$ may proceed either via the release of a chloride and the formation of a vinyl complex or via the protonation of the chloroalkene ligand thus avoiding the formation of the vinyl complex. CPMD simulations were performed in explicit water to provide a microscopic description of the solvent, which is an important aspect for accurate computations of acidity constants. Calculations were performed under periodic boundary conditions using a cubic cell of $\sim$13 Å edge containing the cobalt complex and 57 water molecules. Free energies were obtained by thermodynamic integration. The free energy for the protonation step turned out to be 10 kcal mol^{-1}, much smaller than that estimated from static calculations (30 kcal mol^{-1}). Importantly, AIMD calculations showed that the protonation process is not only thermodynamically favored but also kinetically over the formation of the vinyl complex.

Bühl and coworkers characterized the behavior of the uranyl ion (UO_2^{2+}) in water solution by means of first-principles molecular dynamics [115]. In a first study, water exchange at the hydrated uranyl was investigated. Both the dissociative and associative pathways were evaluated. The associative pathway turned to be the favored one by $\sim$ 4 kcal mol^{-1}, in line with static results but somewhat underestimated compared to experiment [116]. The exchange between the terminal oxo atom and an O atom from bulk water in uranyl hydroxide complexes in basic solution $[UO_2(OH)_4]^{2-}$ was also investigated [117]. Simulations showed that the $[UO_3(OH)_3]^{3-}$ intermediate is obtained via deprotonation of the starting complex. The rate-limiting step turned to be the proton transfer from the hydroxo to the oxo ligand, a process assisted by a solvent water molecule. The overall barrier of 12.5 kcal mol^{-1} was in reasonable agreement with calculated value of 15.2 kcal mol^{-1} obtained from a static model [118], and the agreement is further improved compared to experiments [119].

The hydration/dehydration equilibria of a hydrogentungstate anion, $[WO_3(OH)]^-$; tungstic acid, $[WO_2(OH)_2]$; and protonated tungstic acid, $[WO_2(OH)(H_2O)]^+$ were studied by means of AIMD calculations in a cubic cell box of $\sim$10 Å edge including the metallic species and 29 water molecules by Rodríguez-Fortea, Poblet, and coworkers [120]. The authors showed that increased acidity (decrease of pH) resulted in an expansion of the coordination sphere of the W^{VI}. At the simulated conditions, the tungstic acid did not protonate. Results were not in agreement with continuum methods.

Polyoxometalates (POMs) comprise a class of polynuclear metal–oxygen clusters usually formed by Mo, W, and V and sometimes other metals. With the ambitious and long-term objective of understanding the nucleation mechanism in the formation of the Lindqvist anion $[W_6O_{19}]_{12}^-$, Rodríguez-Fortea, Poblet, and coworkers have investigated dinuclear and trinuclear intermediates that were observed in ESI-MS experiments [121]. Calculations were performed in a periodic cubic box of $\sim$12.6 Å edge including the metal intermediate and about 60 water molecules, depending on the metallic species. The authors observed that the coordination sphere of W^{VI} ions can expand by binding water molecules from the bulk solution. Carbó and coworkers have investigated the behavior of selected Zr-monosubstituted monomeric and dimeric POMs in water solvent at different pHs [122]. Calculations suggested that the Zr metallic centers have a flexible coordination environment, with the tendency to reach coordination numbers greater than six (binding up to three water molecules).

The behavior of other transition metal complexes and ions in aqueous solution has been also investigated. For instance, the solvation of $[Ru(bpy)_3]^{2+}$ in water was evaluated by means of QM/MM-MD calculations in the ground [123] and the triplet excited state [124]. A similar study concerned the $[Fe(bpy)_3]^{2+}$ complex [125]. With the purpose of improving the accuracy of computed Pt NMR chemical shifts, the hydration properties of a set of negatively charged Pt complex ions ($[Pt(X_n)]^{2-}$; X=Cl, Br, and CN) were analyzed by means of AIMD calculations [126]. Hydration studies of square planar Pd^{II}- and Pt^{II}-hydrated complexes revealed interesting behavior in the non-equatorial region, with Pd and Pt behaving differently with respect to the interaction with solvent water molecules [127, 128]. AIMD simulations of *trans*-$PtCl_2(NH_3)(Nglycine)$ in explicit water supported the formation of $Pt\cdots HO$ H-bond interactions between Pt and water molecules in solution [129].

3.2 Reactivity in Nonaqueous Solvents

The application of first-principles molecular dynamics to the study of organometallic systems or homogeneous catalytic processes in nonaqueous solvents is certainly less extended than in water. Nevertheless, a survey of the literature shows that the number of AIMD-based studies in nonaqueous solvents is increasing. In many cases, due to the larger size of the complexes investigated and of the solvent molecules compared to in-water studies, the hybrid QM/MM approach is taken to simulate realistic model systems at a reduced computational cost. Simulations by means of first-principles molecular dynamics on ionic liquids [130] or frustrated Lewis pairs in organic solvents [131] are not covered here.

3.2.1 Ru-Catalyzed Hydrogen Transfer in Methanol

As commented in Sect. 3.1.4, the asymmetric hydrogenation of C=O and C=N bonds is a fundamental transformation in chemistry for the generation of stereogenic centers. Noyori developed Ru-based chiral catalysts [103], whose reaction mechanism has been extensively investigated [132].

Handgraaf and Meijer investigated the reaction mechanism when methanol is the solvent by means of AIMD simulations [133]. Calculations were performed on a model catalyst, $[Ru(C_6H_6)(OCH_2CH_2NH)]$, including 40 molecules of MeOH in a periodic cubic box of ~14 Å edge. Constrained dynamics simulations were performed to estimate the free energy associated with the reaction steps. During the process, a proton and a hydride (H^+, H^-) are formally transferred from methanol to the catalyst in an outer sphere mechanism, i.e., the substrate does not coordinate to the metal. The free energy profile of the process in solution showed a plateau zone associated with the proton transfer from the OH group of the substrate to the N of the amide ligand. This proton transfer results in a terminal oxygen with three lone pairs that was described as a methoxide-like species stabilized by hydrogen bonds from solvent (MeOH) molecules. Calculations performed in gas phase showed that at the transition state, the proton is moving from oxygen (MeOH) to the nitrogen ligand but is not yet fully transferred; on the contrary, in solution, the transfer is complete. Then, the intermediate evolves by transferring the second hydrogen to Ru (formally as H^-), thus converting the methoxide transient species in formaldehyde.

The reverse reaction, i.e., the reduction of formaldehyde to yield methanol, was also studied. It was observed that the hydride moves from Ru to formaldehyde, yielding the methoxide-like species. Such species made a strong hydrogen bond with a MeOH solvent molecule that was also interacting with the N of the amide ligand. Then, a proton transfer from this MeOH solvent molecule to the methoxide-like species took place, converting the former in a methoxide-like species and the latter in methanol. Subsequently, a second solvent molecule donated a proton to the newly formed methoxide molecule. Hence, a chain of proton transfer events was revealed to take place during the process. The simulation was extended to show that eventually the solvent-stabilized methoxide molecule finally exchanges a proton with the Ru complex to form the dehydrogenated Ru complex and methanol.

3.2.2 Ligand Substitution in Pd^0 Complexes in Toluene

Palladium/phosphine systems are among the most used catalytic systems in organic synthesis, as recognized by the assignment of the 2010 Nobel Prize in Chemistry (www.nobelprize.org/nobel_prizes/chemistry/laureates/2010). A debated issue in this field concerns the identification of the catalytically active species, which is crucial for understanding the reactivity of the system. Knowledge of which species are present in solution is a very important step toward this understanding. Pd^0 complex is generated in solution by ligand loss or exchange and may involve participation of solvent molecules. Low coordinated Pd–phosphine complexes are supposed to be more reactive. Pd(phosphine)$_2$ could actually be isolated, whereas the monophosphine species, Pd(phosphine), could not be characterized. Vidossich et al. [134] have used QM/MM-MD simulations to study the speciation of the Pd–phosphine system in explicit toluene. Several ligand exchange processes were evaluated in a periodically repeating box containing Pd^0, tri-phenyl phosphine (s) (PPh$_3$), and around 1,000 toluene molecules.

The behavior of Pd(PPh$_3$) in toluene was addressed by running QM/MM-MD simulations including the metal complex and four toluene molecules in the QM part. A toluene molecule was observed to coordinate to the metal center at the very beginning of the simulation and maintained the coordination to the metal center for the rest of the simulation. Actually, a second toluene molecule was observed to coordinate to the metal center forming $[Pd(PPh_3)(Tol)_2]$. However, the lifetime of this species was limited to few ps. Thus, the monophosphine Pd species is not accessible in toluene, a solvent generally considered as non-coordinating. The species with a single coordinate toluene appears as the most favorable one, but a second solvent molecule may bind the metal center, at least transitorily.

The behavior of $Pd(PPh_3)_2$ in toluene solution was also evaluated. Simulation showed a toluene molecule moving in and out the coordination sphere of Pd during the simulation. Accordingly, the free energy barrier of this process, computed by means of umbrella sampling simulations, was found to be small $(2.4 \pm 0.5$ kcal mol$^{-1})$, with the tricoordinated species lying 1.8 ± 0.6 kcal mol^{-1} higher in energy than the $Pd(PPh_3)_2$ complex. The free energy associated with PPh$_3$ dissociation process from the $[Pd(PPh_3)_2(Tol)]$ complex was estimated to be 17.3 ± 0.6 kcal mol^{-1}. This set of simulations provided evidence for the formation of several solvent adducts such as $[Pd(PPh_3)_2(Tol)]$, $[Pd(PPh_3)(Tol)]$ and $[Pd(PPh_3)(Tol)_2]$, whereas the bare $[Pd(PPh_3)]$ species showed a too limited lifetime to be accessible in solution.

3.2.3 Dynamic Behavior of Agostic Interactions

Agostic interactions are intramolecular interactions between a C–H bond of a ligand and an electron-deficient metal center. These interactions are commonly observed to occupy a vacant coordination site in transition metal complexes, which would otherwise be coordinatively unsaturated. Agostic interactions may be seen as the first stage of C–H bond activation processes. The dynamic behavior of agostic complexes was investigated by QM/MM-MD simulations by Ortuño et al. [135]. The objective of this study was to reveal whether the C–H is kept firmly bonded to the metal center or rather fluxionality among equivalent C–H bonds would take place. The authors selected a set of three-coordinated T-shaped Pt^{II} complexes showing diverse agostic geometries. In the three Pt^{II} complexes considered, agostic interactions may involve C–H bonds at β (complex Pt_1), δ (complex Pt_2), and ζ positions (complex Pt_3, Fig. 8). The organometallic complexes were described at the quantum mechanical level, whereas the ~ 1,000 dichloromethane solvent molecules and the $[SbF_6]^-$ counterion were described by means of molecular mechanics. Simulations were run for 15 ps for each of the complexes at 300 K. The results show quite different behavior in solution for each complex.

In complex Pt_1 the agostic interaction is maintained throughout the simulation. The average Pt$\cdots$C and Pt$\cdots$H distances were comparable to those obtained from geometry optimization. Simulation of complex Pt_2 showed that the methyl groups of the *tert*-butyl substituent interchanged interaction with the metal center, providing an atomistic picture on how the averaging of the NMR signals of the three methyl groups takes place. Simulation of complex Pt_3, in which the C–H bond involved in the agostic interaction is six bonds away from the metal center, showed an easy rotation of the CH$_3$ groups of the isopropyl substituent. It was also observed that for short time intervals (few ps), the agostic interaction may break. Thus, these

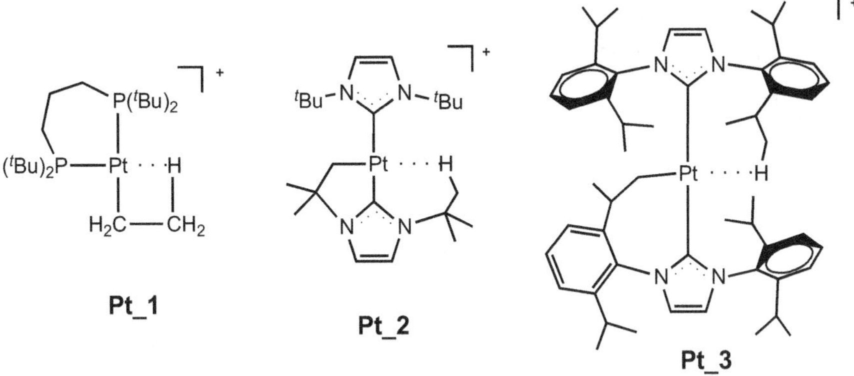

Fig. 8 Schematic representation of the three Pt^{II} complexes featuring agostic interactions studied by Ortuño et al

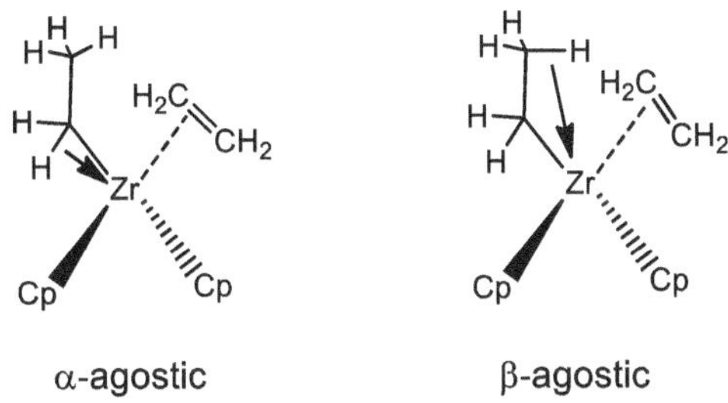

Fig. 9 Schematic representation of α- and β-agostic zirconocene complexes studied by Rowley and Woo [135]

set of simulations were able to reveal a range of dynamical behaviors of Pt···H–C agostic interactions.

Group IV metallocene catalysts are of major importance in olefin polymerization chemistry. Rowley and Woo [136] studied the dynamic behavior of agostic interactions in an archetypal zirconocene–olefin complex, $[(Cp)_2Zr(C_2H_5)(C_2H_4)]^+$. The authors were interested in exploring how the counterion, $[CH_3B(C_6F_5)_3]^-$, affects the reactivity of the complex. More specifically, the counterion effect on the relative stability and the free energy barrier for the interconversion process between the α- and β-agostic interactions were addressed (Fig. 9). Molecular dynamics is indeed an appropriate approach to the modeling of ion pairs in solution. QM/MM-MD simulations were performed at 300 K in explicit pentane solvent. The metallocene was described at QM level, whereas the counterion and the solvent molecules were represented at the MM level. The interconversion process between the α- and β-agostic complexes was studied in the presence and absence of the counterion. Interestingly, simulations of the complex in the presence of the counterion showed spontaneous isomerization from the β-agostic to the α-agostic configuration. Conversely, static calculations predict the former to be more stable by

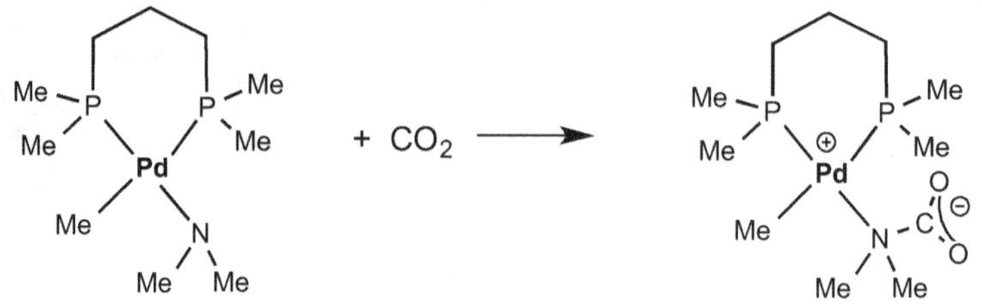

Fig. 10 Reaction step studied for CO_2 activation by a Pd complex in CO_2 under supercritical conditions

2.7 kcal mol^{-1}. The free energy profile of this process was calculated using umbrella sampling simulations. The barrier was found to be 2.8 kcal mol^{-1}, with the β-agostic species 1.6 kcal mol^{-1} lower in energy. Calculations including the counterion resulted in a barrier of 1.9 kcal mol^{-1}, with the β-agostic species 1.0 kcal mol^{-1} more stable than the α-agostic species. The authors attributed the stabilization in the presence of the counterion to the more favorable dipole–counteranion interaction in the more polar α-agostic configuration.

3.2.4 Addition of CO_2 to a Pd Complex in CO_2

Growing interest is shown toward catalysts capable of activating small molecules such as carbon dioxide [137]. Kirchner and coworkers [138] studied the addition of carbon dioxide to a Pd complex with an amide ligand, leading to the formation of a carbamate–palladium complex. More specifically, the reaction step corresponding to the formation of the C–N bond between the amide ligand and the incoming CO_2 was investigated by means of metadynamics simulations (Fig. 10). The model system included the Pd complex and 60 carbon dioxide molecules in a periodically repeating simulation box (supercritical CO_2 conditions were considered). In addition, molecular dynamics simulations in gas phase were also performed.

Comparison of the free energy surface for the process in gas phase and supercritical CO_2 revealed an increase in the free energy barrier for the bulk conditions (6.4 ± 0.5 kcal mol^{-1} in vacuum and 10.7 ± 0.8 kcal mol^{-1} in supercritical CO_2). Analysis of the results for the reverse reaction showed that the change in entropy associated with the dissociation process is less important (increase less) in supercritical CO_2 than in gas phase. This leads to a higher stabilization of the palladium–carbon dioxide complex, pointing out that solvent effects have an important contribution to the stabilization of the palladium–carbon dioxide adduct.

4 Conclusions and Remarks

We have reviewed the application of explicit solvent first-principles molecular dynamics simulation to the study of the reactivity of organometallic systems. We have pointed out how the mechanistic insight provided by this approach may turn out to be qualitatively and quantitatively different from results of in vacuo and continuum modeling. Indeed, as we have shown, the presence of explicit solvent molecules can have a considerable impact when processes such as proton transfer or the formation of charged or unsaturated species are investigated. This is certainly well established for reactions in water, because of its peculiar properties (high dielectric, hydrogen bonding, acid–base, coordinative), and the early application of the AIMD methodology indeed addressed reactivity in this medium. However, even for nonaqueous solvents, the description of organometallic reactivity may improve when extended systems are considered.

The AIMD methodology is nowadays well established, and in this review, we put the emphasis on its strengths. However, it does also have some limitations. One is certainly the computational cost associated with an increased number of atoms, which limits the time scales that can be investigated. Advanced simulation techniques have been developed in order to promote the occurrence of rare events on the time scale currently accessible (hundreds of ps). Yet, the computational cost remains high, and the exploration of alternative reaction paths or different electronic surfaces remains a challenge. The hybrid quantum mechanical/molecular mechanical approach is an effective way to address larger complexes, especially the ones with bulk ligands that are of high interest for selective transformations.

Having pointed out pros and cons of explicit solvent AIMD simulations applied to organometallic reactivity, it is likely that this approach will continue to coexist and complement the computational chemistry approach based on the potential energy exploration of reduced model systems in the near future. On the long term, however, we would like to push the modeling to the point in which reactants are mixed in a virtual reactor and their transformation to products registered (see, e.g., [139] for a recent report showing that the approach is indeed within our reach). Here, the use of realistic models will certainly be advantageous, as no a priori assumption on which species are relevant for reactivity would be introduced.

Acknowledgments The authors warmly thank all collaborators with whom they share the interest in the application of first-principles simulations to the study of homogeneous catalysis. We specially thank former members of the group Aleix Comas-Vives, Gábor Kovács, and Manuel A. Ortuño and long collaborators as András Stirling, or more recent Nisanth N. Nair, who contributed to some applications presented in this review. The authors thankfully acknowledge the computer resources, technical expertise, and assistance provided by the Barcelona Supercomputing Center (Centro Nacional de Supercomputación). We gratefully acknowledge financial support from Spanish MINECO (CTQ2014-54071-P).

References

1. Cundari TR (ed) (2001) Computational organometallic chemistry. Marcel Dekker, New York
2. Maseras F, Lledós A (eds) (2002) Computational modeling of homogeneous catalysis. Kluwer, Dordrecht
3. Morokuma K, Musaev DG (eds) (2008) Computational modeling for homogeneous and enzymatic catalysis. Wiley, Weinheim

4. Lin Z (2010) Acc Chem Res 43:602
5. Tsang ASK, Sanhueza IA, Schoenebeck F (2014) Chem Eur J 20:1
6. Thiel W (2014) Angew Chem Int Ed 53:2
7. Davidson ER (2000) Chem Rev 100:351
8. Bell AT, Head-Gordon M (2011) Annu Rev Chem Biomol Eng 2:453
9. Cramer CJ, Truhlar DG (1999) Chem Rev 99:2161
10. Tomasi J, Mennucci B, Cammi R (2005) Chem Rev 105:2999
11. Curutchet C, Cramer CJ, Truhlar DG, Ruiz-López MF, Rinaldi D, Orozco M, Luque FJ (2003) J Comput Chem 24:284
12. Sunoj RB, Anand M (2012) Phys Chem Chem Phys 14:12715
13. Ortuño MA, Lledós A, Maseras F, Ujaque G (2014) ChemCatChem 6:3132
14. Diez J, Gimeno J, Lledós A, Suarez FJ, Vicent C (2012) ACS Catal 2:2087
15. Rodríguez-Santiago L, Alí-Torres J, Vidossich P, Sodupe M (2015) Phys Chem Chem Phys 17:13582
16. Van Speybroeck V, Meier RJ (2003) Chem Soc Rev 32:151
17. Margl P, Ziegler T, Blöchl PE (1995) J Am Chem Soc 117:12625
18. De Angelis F, Fantacci S, Sgamellotti A (2006) Coord Chem Rev 250:1497
19. Allen MP, Tildesley DJ (1989) Computer Simulation of Liquids. Oxford University Press, Oxford
20. Frenkel D, Smit B (2001) Understanding molecular simulation: from algorithms to applications. Academic, San Diego
21. Verlet L (1967) Phys Rev 159:98
22. Nose S (1984) Mol Phys 52:255
23. Bussi G, Donadio D, Parrinello M (2007) J Chem Phys 126:014101
24. Martyna GJ, Tobias DJ, Klein ML (1994) J Chem Phys 101:4177
25. Marx D, Hutter J (2009) Ab initio molecular dynamics: basic theory and advanced methods. Cambridge University Press, Cambridge
26. Car R, Parrinello M (1985) Phys Rev Lett 55:2471
27. Hohenberg P, Kohn W (1964) Phys Rev B 136:B864
28. Cramer CJ, Truhlar DG (2009) Phys Chem Chem Phys 11:10757
29. Neese F (2009) Coord Chem Rev 253:526
30. Becke AD (2014) J Chem Phys 140:18A301
31. Kohn W, Sham LJ (1965) Phys Rev 140:1133
32. Parr RG, Yang W (1994) Density-functional theory of atoms and molecules. Oxford University Press, New York
33. Koch W, Holthausen MC (2001) A chemist's guide to density functional theory. Wiley, Weinheim
34. Cohen AJ, Mori-Sanchez P, Yang W (2012) Chem Rev 112:289
35. Zhao Y, Truhlar DG (2008) Theor Chem Acc 120:215
36. Grimme S (2006) J Comput Chem 27:1787
37. Grimme S, Antony J, Ehrlich S, Krieg H (2010) J Chem Phys 132:154104
38. Kulik HJ, Cococcioni M, Scherlis DA, Marzari N (2006) Phys Rev Lett 97:103001
39. VandeVondele J, Sprik MA (2005) Phys Chem Chem Phys 7:1363
40. Marques MAL, Gross EKU (2004) Annu Rev Phys Chem 55:427
41. Neese F (2006) J Biol Inorg Chem 11:702
42. Goedecker S (1999) Rev Mod Phys 71:1085
43. Sulpizi M, Raugei S, VandeVondele J, Carloni P, Sprik M (2007) J Phys Chem B 111:3969
44. Weber V, Bekas C, Laino T, Curioni A, Bertsch A, Futral S (2014) In: Parallel and distributed processing symposium, 2014 IEEE. 28th international 735
45. Warshel A, Levitt M (1976) J Mol Biol 103:227
46. Field MJ, Bash PA, Karplus M (1990) J Comput Chem 11:700
47. Singh UC, Kollman PA (1986) J Comput Chem 7:718
48. Senn HM, Thiel W (2009) Angew Chem Int Ed 48:1198

49. Rovira C (2013) WIREs Comput Mol Sci 3:393
50. Bo C, Maseras F (2008) Dalton Trans 2911
51. Ananikov VP, Musaev DG, Morokuma K (2010) J Mol Catal A Chem 324:104
52. Sameera WMC, Maseras F (2012) WIREs Comput Mol Sci 2:375
53. Woo TK, Margl PM, Deng L, Cavallo L, Ziegler T (1999) Catal Today 50:479
54. Woo TK, Blöchl PE, Ziegler T (2000) J Mol Struct (teochem) 506:313–334
55. Guidoni L, Maurer P, Piana S, Rothlisberger U (2002) Quant Struct-Act Relat 21:119
56. Laino T, Mohamed F, Laio A, Parrinello M (2005) J Chem Theory Comput 1:1176
57. Laino T, Mohamed F, Laio A, Parrinello M (2006) J Chem Theory Comput 2:1370
58. Rowley CN, Roux B (2012) J Chem Theory Comput 8:3526
59. Nielsen SO, Bulo RE, Moore PB, Ensing B (2010) Phys Chem Chem Phys 12:12401
60. Bernstein N, Várnai C, Solt I, Winfield SA, Payne MC, Simon I, Fuxreiter M, Csányi G (2012) Phys Chem Chem Phys 14:646
61. Steinfeld JI, Francisco JF, Hase WL (1998) Chemical kinetics and dynamics. Prentice Hall, Upper Saddle River
62. Pohorille A, Chipot C (eds) (2007) Free Energy Calculations: Theory and Applications in Chemistry and Biology. Springer, Berlin
63. Kirkwood JG (1935) J Chem Phys 3:300
64. Laio A, Parrinello M (2002) Proc Natl Acad Sci U S A 99:12562
65. Barducci A, Bonomi M, Parrinello M (2011) WIREs Comput Mol Sci 1:826
66. Ensing B, De Vivo M, Liu Z, Moore P, Klein ML (2006) Acc Chem Res 39:73
67. Carter EA, Ciccotti G, Hynes JT, Kapral R (1989) Chem Phys Lett 156:472
68. Sprik M, Ciccotti G (1998) J Chem Phys 109:7737
69. Laio A, Gervasio FL (2008) Rep Prog Phys 71:126601
70. Ensing B, Laio A, Parrinello M, Klein ML (2005) J Phys Chem B 109:6676
71. Zheng S, Pfaendtner J (2014) Mol Simul 41:55
72. Laio A, Rodriguez-Fortea A, Gervasio FL, Ceccarelli M, Parrinello M (2005) J Phys Chem B 109:6714
73. Iannuzzi M, Laio A, Parrinello M (2003) Phys Rev Lett 90:238302
74. Bolhuis PG, Chandler D, Dellago C, Geissler PL (2002) Annu Rev Phys Chem 53:291
75. Hutter J, Curioni A (2005) ChemPhysChem 6:1788
76. VandeVondele J, Krack M, Mohamed F, Parrinello M, Chassaing T, Hutter J (2005) Comput Phys Commun 167:103
77. Hassanali AA, Cuny J, Verdolino V, Parrinello M (2014) Phil Trans R Soc A 372:20120482
78. Haber F, Weiss J (1934) Proc R Soc Lond 147:332
79. Bray WC, Gorin MH (1932) J Am Chem Soc 54:2124
80. Ensing B, Buda F, Blöchl PE, Baerends EJ (2001) Angew Chem Int Ed 40:2893
81. Ensing B, Buda F, Blöchl PE, Baerends EJ (2002) Phys Chem Chem Phys 2:3619
82. Yamamoto N, Koga N, Nagaoka M (2012) J Phys Chem B 116:14178
83. Ensing B, Buda F, Gribnau MCM, Baerends EJ (2004) J Am Chem Soc 126:4355
84. Shaik S, Kumar D, de Visser SP, Altun A, Thiel W (2005) Chem Rev 105:2279
85. Louwerse MJ, Vassilev P, Baerends EJ (2008) J Phys Chem A 112:1000
86. Bernasconi L, Baerends EJ (2013) J Am Chem Soc 135:8857
87. Jira R (2009) Angew Chem Int Ed 48:9034
88. Keith JA, Henry PM (2009) Angew Chem Int Ed 48:9038
89. Stirling A, Nair NN, Lledós A, Ujaque G (2014) Chem Soc Rev 43:4940
90. Kovács G, Stirling A, Lledós A, Ujaque G (2012) Chem Eur J 18:5612
91. Nair NN (2011) J Phys Chem B 115:2312
92. Imandi V, Kunnikuruvan S, Nair NN (2013) Chem Eur J 19:4724
93. Bäckvall JE, Akermark B, Ljunggren SO (1979) J Am Chem Soc 101:2411
94. Beyramabadi SA, Eshtiagh-Hosseini H, Housaindokht MR, Morsali A (2009) J Mol Struct (THEOCHEM) 903:108
95. Henry PM (1973) J Org Chem 38:2415

96. Heck RF (1968) Hercules Chem 57:12
97. Keith JA, Nielsen RJ, Oxgaard J, Goddard WA (2006) J Am Chem Soc 128:3132
98. Imandi V, Nair NN (2015) J Phys Chem B 119:11176
99. Carloni P, Sprik M, Andreoni W (2000) J Phys Chem B 104:823
100. Bose RN, Cornelius RD, Viola RE (1984) J Am Chem Soc 106:3336
101. Lau JKC, Ensing B (2010) Phys Chem Chem Phys 12:10348
102. Melchior A, Tolazzi M, Martínez JM, Pappalardo RR, Sánchez-Marcos E (2015) J Chem Theory Comput 11:1735
103. Noyori R, Hashiguchi S (1997) Acc Chem Res 30:97
104. Pavlova A, Meijer EJ (2012) ChemPhysChem 13:3492
105. Bandaru S, English NJ, MacElroy JMD (2014) J Comput Chem 35:683
106. Li P, Henkelman G, Keith JA, Johnson JK (2014) J Phys Chem C 118:21385
107. Lewis NS, Nocera DG (2008) Proc Natl Acad Sci U S A 103:15729
108. Kohl SW, Winer L, Schwartsburd L, Konstantinovski L, Shimon LJW, Ben-David Y, Iron MA, Milstein D (2009) Science 324:74
109. Ma C, Piccinin S, Fabris S (2012) ACS Catal 2:1500
110. Vallés-Pardo JL, Guijt MC, Iannuzzi M, Joya KS, de Groot HJM, Buda F (2012) ChemPhysChem 13:140
111. Vallés-Pardo JL, de Groot HJM, Buda F (2012) Phys Chem Chem Phys 14:15502.
112. Shilov AE, Shul'pin GB (1997) Chem Rev 97:2879
113. Vidossich P, Ujaque G, Lledos A (2012) Chem Commun 48:1979
114. Bühl M, Golubnychiy V (2007) Organometallics 26:6218
115. Bühl M, Wipff G (2011) ChemPhysChem 12:3095
116. Bühl M, Kabrede H (2006) Inorg Chem 45:3834
117. Bühl M, Schreckenbach G (2010) Inorg Chem 49:3821
118. Shamov GA, Schreckenbach G (2008) J Am Chem Soc 130:13735
119. Szabo Z, Grenthe I (2007) Inorg Chem 46:9372
120. Rodríguez-Fortea A, Vilà-Nadal L, Poblet JM (2008) Inorg Chem 47:7745
121. Vilà-Nadal L, Rodríguez-Fortea A, Poblet JM (2009) Eur J Inorg Chem 5125
122. Jiménez-Lozano P, Carbó JJ, Chaumont A, Poblet JM, Rodríguez-Fortea A, Wipff G (2014) Inorg Chem 53:778
123. Moret ME, Tavernelli I, Rothlisberger U (2009) J Phys Chem B 113:7737
124. Moret ME, Tavernelli I, Chergui M, Rothlisberger U (2010) Chem Eur J 16:5889
125. Daku LML, Hauser A (2010) J Phys Chem Lett 1:1830
126. Truflandier LA, Autschbach J (2010) J Am Chem Soc 132:3472
127. Beret EC, Pappalardo RR, Doltsinis NL, Marx D, Sánchez-Marcos E (2008) ChemPhysChem 9:237
128. Beret EC, Martínez JM, Pappalardo RR, Sánchez-Marcos E, Doltsinis NL, Marx D (2008) J Chem Theory Comput 4:2108
129. Vidossich P, Ortuño M, Ujaque G, Lledós A (2011) ChemPhysChem 12:1666
130. Zahn S, Brehm M, Brüssel M, Hollóczki O, Kohagen M, Lehmann S, Malberg F, Pensado AS, Schöppke M, Weber H, Kirchner B (2014) J Mol Liq 192:71
131. Pu M, Privalov T (2015) Isr J Chem 55:179
132. Dub PA, Henson NJ, Martin RL, Gordon JC (2014) J Am Chem Soc 136:3505
133. Handgraaf JW, Meijer EJ (2007) J Am Chem Soc 129:3099
134. Vidossich P, Ujaque G, Lledós A (2014) Chem Commun 50:661
135. Ortuño MA, Vidossich P, Ujaque G, Conejero S, Lledós A (2013) Dalton Trans 42:12165
136. Rowley CN, Woo TK (2010) Organometallics 30:2071
137. Aresta M (2010) Carbon dioxide as chemical feedstock. Wiley, Weinheim
138. Brüssel M, di Dio PJ, Muñiz K, Kirchner B (2011) Int J Mol Sci 12:1389
139. Wang LP, Titov A, McGibbon R, Liu F, Pande VS, Martínez TJ (2014) Nat Chem 6:1044

Struct Bond (2016) 167: 107–138
DOI: 10.1007/430_2014_151
© Springer-Verlag Berlin Heidelberg 2014
Published online: 24 July 2014

Computation of Excited States of Transition Metal Complexes

Nuno M. S. Almeida, Russell G. McKinlay, and Martin J. Paterson

Abstract In this review we discuss the theory and application of methods of excited state quantum chemistry to excited states of transition metal complexes. We review important works in the field and, in more detail, discuss our own studies of electronic spectroscopy and reactive photochemistry. These include binary metal carbonyl photodissociation and subsequent non-adiabatic relaxation, Jahn–Teller and pseudo-Jahn–Teller effects, photoisomerization of transition metal complexes, and coupled cluster response theory for electronic spectroscopy. We aim to give the general reader an idea of what is possible from modern state-of-the-art computational techniques applied to transition metal systems.

Keywords Computational chemistry · Excited states · Photochemistry · Electronic spectroscopy · Non-adiabatic chemistry · Jahn-Teller theory

Contents

1 Introduction and Theoretical Background

The application of computational and theoretical methods to organometallic chemistry has been the subject of many surveys during the past few years. Here we focus on an area that has seen much less activity: excited state computation of inorganic

N.M.S. Almeida, R.G. McKinlay, and M.J. Paterson (✉)
Institute of Chemical Sciences, School of Engineering and Physical Sciences, Heriot-Watt University, Edinburgh EH14 4AS, UK
e-mail: m.j.paterson@hw.ac.uk

and organometallic complexes. This includes aspects of both theoretical spectroscopy and reactive photochemistry. Nowadays, with the current evolution of computer hardware, and the development of different theories of computational chemistry, scientists can improve and explore deeper to get a better understanding of these types of systems than was possible until relatively recently. This evolution of computational chemistry in terms of both software and hardware means that there are now several methodologies that can be employed when studying these difficult and demanding problems.

Computational methods such as density functional theory (DFT) [1–3], Moller–Plesset perturbation theory (MP) [4, 5], and coupled-cluster (CC) theory [6–8] are common computational methods used to calculate ground state properties with quantum chemistry. These methods are "black-box" in the sense that the system can be analysed solely in terms of giving an input structure and a level of theory. When a problem is formulated, the choice of which method to choose comes down to a balance between levels of accuracy required versus computational expense. Even for ground state problems black-box methods may not be applicable in all cases. For example, in cases of near or actual degeneracy then any method based on a single determinant as a reference will be invalid. Such things are far more common when transition metals are involved.

When one is focussing on excited electronic states, DFT, MP, and CC methods are inappropriate because they only allow calculation of the lowest state of a given symmetry. They can however be extended into the excited domain by the use of response theory and its variants. Here one includes the effects of time-dependent perturbations (e.g., the electric field of light) on the fundamental quantities in the theory: the effect on density in DFT, effect on cluster amplitudes in CC theory. A very popular method is the so-called time-dependent density functional theory (TD-DFT), which allows straightforward (and inexpensive) computation of electronically excited states when a standard Kohn–Sham ground state treatment is valid. It is worth noting that if the ground state has near or actual degeneracies then this method fails completely as the ground state must be describable within the standard framework. Naturally these response-based methods are excellent for simulating electronic spectroscopy around a stable ground state minimum. The formulation allowing us to describe relaxation in the excited state via gradients of the excited potential energy surfaces is more complex but recent advances allow them to be possible. Very recently TD-DFT excited state gradients became available as a standard tool. Organic molecules have seen much application of TD-DFT, and coupled cluster response, in both the equation of motion (EOM) formalism and the linear response (LR) one. One enticing advantage of the coupled cluster response is that a systematic improvement in excited state properties can be obtained by going to a calculation with the next level of CC reference. It can formally be shown that for a given one-electron basis set there is a hierarchy of methods correct to a given order in the fluctuation potential (the difference between the mean-field Hartree–Fock potential and the exact Coulomb potential). For example, CCS, CC2, CCSD, CC3, CCSDT, in which excited states obtained from respective response functions give excitation energies correct to order 1, 2, 3, 4, etc. [9, 10]. See Fig. 1 for a graphical representation. CC2 (and CC3) are approximations

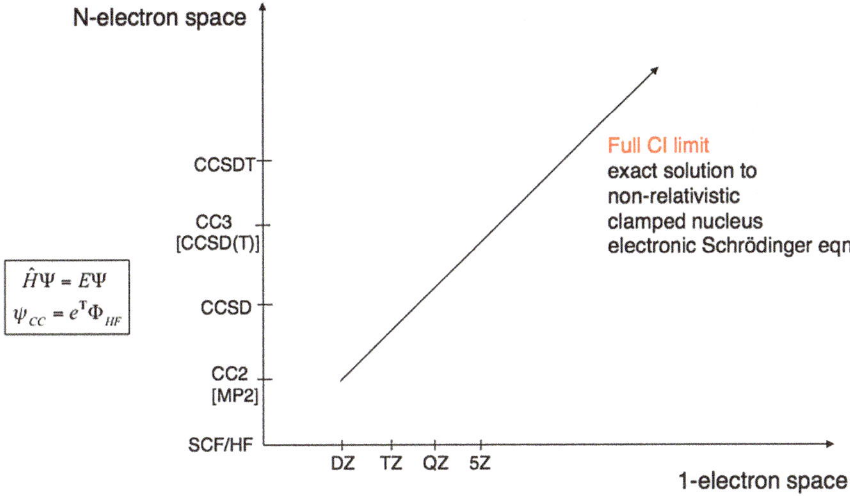

Fig. 1 Graphical illustration of the convergence of properties computed from CC response functions in both the one- and N-particle spaces. Computational cost increases along both axes as well

to CCSD (and CCSDT, respectively) in which the doubles for CC2 and triples for CC3 are approximated with expressions developed via perturbation theory.

The price to be paid is the computational scaling with each method. Thus CCSD scales as N^6, while CCSDT scales as N^8, where N is representative of the system size. Thus, doubling the size of a system leads to a 64-fold increase in the cost of the CCSD calculation. Therefore the range of applicability of such methods is limited to quite small systems. Only very limited application to inorganic and organometallic systems has been seen. However, such correlated ab initio methods are invaluable for providing benchmarks, and being able to treat all types of states in a balanced manner. Recently the first applications of CC response theory to inorganic systems have begun to appear. However it is worth noting that application of correlated response methods to transition metal containing systems can be a very demanding and difficult task, with more pitfalls than for organic systems [11]. Some of the recent applications are discussed below. Despite the heavy cost of the CC response hierarchy there are some advantages in that the cost of determining the response functions is quite low compared to the ground state CC wavefunction, and thus many states can be calculated with only very small overhead. Also, the theoretical machinery means that non-linear spectroscopies, such as two-photon absorption (TPA), can be calculated as straightforwardly as for one-photon absorption (via the quadratic response function instead of the linear response function). We mention TPA in relation to $Fe(CO)_5$ ((**1**) in Scheme 1) later in the review.

As discussed above when the ground state cannot be represented by a single electronic configuration then the response machinery for DFT and CC becomes

N.M.S. Almeida et al.

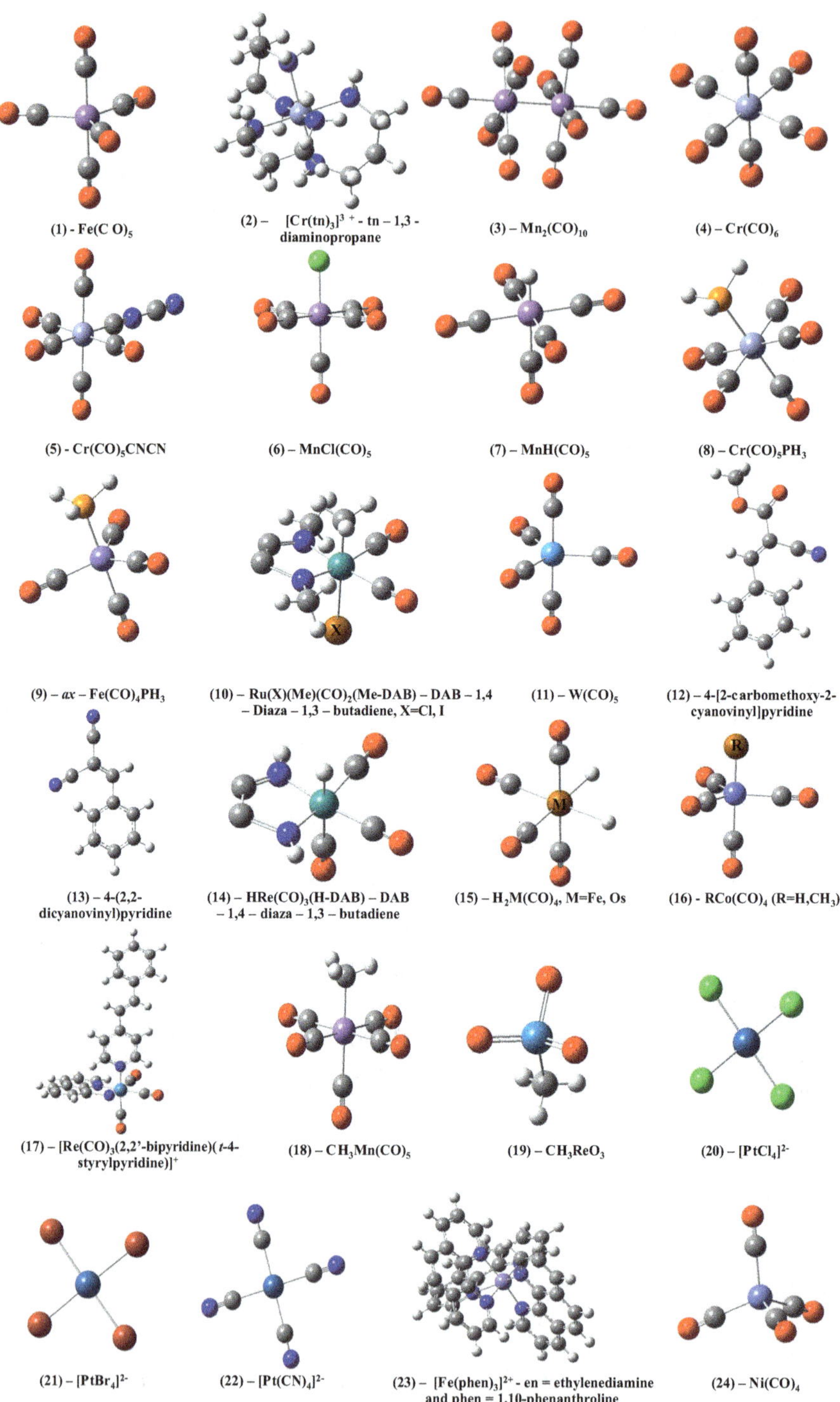

Scheme 1 (continued)

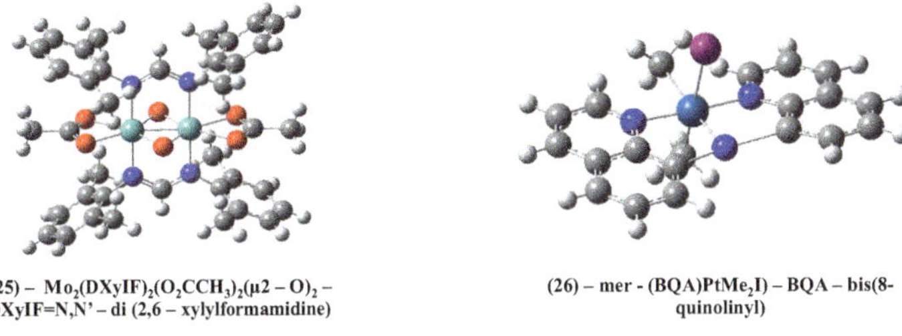

Scheme 1 Computational studies of the photochemistry of these inorganic molecules are reviewed

inapplicable. Now one must describe both the ground and excited electronic states via appropriate wavefunctions. This allows us to determine excited state reaction paths and to investigate regions of strong non-adiabatic couplings (i.e., degeneracies, particularly excited states(s) with the ground state).

Configuration interaction (CI) involves expanding the wavefunction in terms of configurations built from a reference by "exciting" i.e., substituting electrons from occupied to virtual orbitals, from, for example, a Hartree–Fock calculation. Similar to CC theory (above) CI theory involves amplitudes determining the highest number of substituted orbitals allowed in the expansion. Here S means singles, D means doubles, T means triples, and so on. The difference between the two is that the CC expansion is non-linear (exponential) and includes the so-called disconnected terms (e.g., quadruples as products of doubles, etc.), whereas CI is a linear expansion method. Thus for ground states CC is generally preferred and leads to orders of magnitude in increased accuracy. If up to full N-electron substitutions are considered, then we generate a full CI wavefunction. This is a numerically exact solution to the many-electron problem in a given one-particle basis (i.e., top of vertical axis in Fig. 1). The scaling of full CI is factorial and has only ever been applied to small molecules, for example the reaction $F + H_2 \rightarrow HF + H$, [12]. Full CI does have several desirable properties for excited state computation though, namely that as a variational method the ground and all excited state eigenvalues are upper bounds to the state energies. Moreover the linear expansion generates very flexible wavefunctions capable of, in principle, describing any possible electronic state at any possible geometry, including degeneracies. Thus methods incorporating full CI (in a subset of the orbitals) have been developed for such purposes: namely complete-active space self-consistent field (CASSCF theory) [13].

CASSCF is a variant of multi-configuration self-consistent field (MCSCF) theory. This means that in addition to the CI expansion coefficients being variationally optimized, the orbitals determining the expansion are also variationally optimized. Thus, the linear combination of atomic orbitals (LCAO) coefficients defining the orbitals is simultaneously optimized. If one starts with, for example, Hartree–Fock orbitals, then after the MCSCF wavefunction is optimized the orbitals will (often) be quite different. MCSCF wavefunctions thus contain the optimum orbitals for the given CI expansion. CASSCF involves choosing a subset of orbitals and a number of electrons, and performing a full CI calculation in this space, plus optimization of the orbitals [14]. CASSCF is capable of providing qualitative wavefunctions when the electronic structure requires more than a single configuration (called static correlation), and this includes cases where a linear combination of configurations is required to correctly represent a spin-state (spin-correlation, e.g., two configurations for an open-shell singlet state). CASSCF formulates the model to solve a problem in terms of orbitals, rather than configurations (which is the case for general MCSCF). There is no unique correct choice of active space, but rather the user has to choose orbitals relevant to the problem under consideration, alongside the cost of the CI expansion. The current limit is around 14 electrons in 14 orbitals for Gaussian09 [15]. The initial orbitals (the 'guess' set) often determine whether a wavefunction will converge or not. One ideally wants the active space to include all strongly correlated orbitals (i.e., those with occupations substantially different from zero for Occupied or two for virtual). For transition metal containing systems this is often the key aspect of using the method. For example, if one wishes to consider and compare both ligand field (LF) and charge transfer (CT) states, then one needs both metal-centred and ligand-delocalized orbitals. This gives very large CI expansions but has been very successful in modeling electronic spectra (several examples are discussed below). For reactive photochemistry generally the active spaces will be a bit smaller as one needs to compute energy gradients and non-adiabatic couplings, etc. Examples of choosing active spaces for this are also considered below. If several electronic states are close in energy, then the state-averaged (SA) CASSCF procedure is best employed. Here a weighted average of the orbitals is optimized. This avoids the problem of root flipping where an excited state drops below a lower state due to variational optimization, and the method then focuses on the upper state, until that drops below, then switches state again, etc. ad infinitum. CASSCF is available in a number of commercial codes such as Gaussian [15], Molcas [16], Columbus [17–21], and Gamess UK [22].

While CASSCF generates good qualitative wavefunctions (and occasionally quantitative ones for small systems) the energetics emerging are still far from ideal due to the neglect of dynamic correlation. This is caused by the instantaneous repulsion of pairs of electrons and is the correlation that MP, CC, and DFT are designed to account for. CASSCF can be extended with further CI on the fixed CASSCF orbitals to give multi-reference (MRCI) methods. These are very expensive and haven't seen much use with inorganic systems. By far the most popular alternative is the relatively computationally inexpensive step of applying second-order perturbation theory to a CASSCF wavefunction. There is no unique way to define this but the variant CASPT2 [22–24] has emerged as the most robust and accurate formalism that has seen general use. CASPT2 has seen extensive use by a few specialized groups working in transition metal photochemistry and spectroscopy and several examples are discussed below.

With a suitable wavefunction one can investigate regions of potential energy surfaces involving non-adiabatic vibronic coupling. Conical intersections have been found to be crucial in the mechanistic explanation of the photochemical

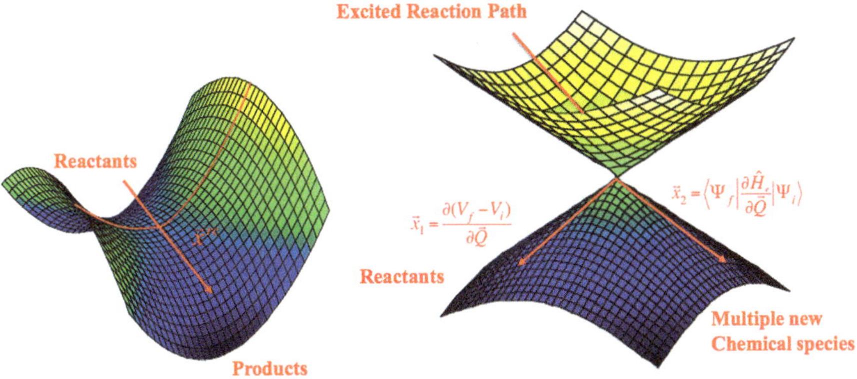

Fig. 2 Comparison of reaction paths through a transition state vs. a conical intersection

behaviour of a variety of different organic and inorganic systems. They have been a subject of an increasing number of studies in the past two decades. They occur at degeneracies between molecular potential energy surfaces and provide radiationless decay channels that influence the formation of different photoproducts. Two states are degenerate when they have the same electronic energy. In high-symmetry molecular systems degeneracy implies the Jahn–Teller effect which are simply conical intersections imposed by symmetry (the degenerate states are simply any pair of linearly independent combinations of the components of the degenerate electronic state). Studies on the mechanisms of such photoreactions and the molecular structure causing radiationless decay are now routine for organic photoreactions. It is worth noting the differences in photochemical reactions occurring via a conical intersection with thermal reactions confined to a single potential energy surface. A transition state connects the reactants with a single product in a single reaction path (motion along transition vector $\vec{x}^{\,rc}$ (Fig. 2). Conical intersection can connect reactants with two or more photoproducts via many reaction paths (branching of the reaction path in the plane $\vec{x}_1$ and $\vec{x}_2$).

The $\vec{x}_1$ and $\vec{x}_2$ vectors presented in Fig. 2 are known as the gradient difference and derivative coupling vectors, which are special internal coordinates that lift the degeneracy to first order in nuclear motion. The remaining 3N-8 internal coordinates do not lift the degeneracy at first-order, and they span the intersection space, which is a hyperline consisting of an infinite number of conical intersection points (known as the seam).

The position and local topology of a conical intersection is crucial in determining the photochemical reactivity of a molecule. A peaked conical intersection occurs when both PES have orthogonal gradients near the intersection (such as that in Fig. 2). Such topology leads to very efficient transitions from the excited state to the ground state. An example of such a conical intersection will be presented below for several systems. A sloped conical intersection occurs when both PES have almost parallel gradients near the intersection and this topology is discussed for $[Cr(tn)_3]^{3+}$ ((2) in Scheme 1) above. When two PESs of different spin states cross such as in an intersystem crossing the $\vec{x}_2$ vector is zero by symmetry and the surfaces cross in a seam of dimension 3N-7. Conical intersections are often called 'photochemical funnels' in organic photochemistry as they work as very efficient mechanisms for a system passing from an upper (excited) to a lower potential surface. The system can cross to the lower surface within a few (possibly only one) vibrational period(s). This is especially important in, for example, photo-stabilization mechanisms. In order to evaluate the efficiency of going through a conical intersection dynamics calculations are required as the system can cross at any point on the seam of intersection. Different methods are available for this task, each with their own advantages and disadvantages (primarily computational cost and scaling) from classical trajectory to quantum wavepacket simulations. Quali-tative information can be retrieved for larger systems with classical trajectory simulations, and for relatively small systems quantum wavepacket calculations may be employed with a suitable model Hamiltonian (e.g., linear vibronic cou-pling). As the Born–Oppenheimer approximation breaks down at the point of intersection care must be taken when performing dynamics calculations in these cases. For further information on this point, see [25, 26].

In all these cases so far we assume that the system is calculated in isolation in the gas phase. This is a standard theoretical starting point for many studies. For many fundamental photochemical reactions it is possible to study the isolated gas phase and such experiments are crucial in connecting with the theoretical treatments discussed. However, there are many cases where the system cannot be prepared in the gas phase or where the reactive chemistry does not occur in the gas phase, for example in biochemistry with large pharmaceutically important molecules. In these cases the effect of the environment must also be considered, for example through solvation models. This can be accomplished in a number of possible ways within the majority of commercial computational chemistry codes. It is important to state this is not trivial when considering how to go about this task. One can use a polarized continuum model that places the system in a solvent cavity based on overlapping spheres characterized by the dielectric constant or electrostatic poten-tial of the solvent. No specific solvent molecules are described in this approach. Specific solvent molecules can be specified around the system of interest if the exact interaction with solvent molecules is important leading to the so-called semi-continuum models, with a corresponding increase in computational cost. One can also split the central system and solvent sphere, most popularly QM/MM calcula-tions where a DFT or ab initio method like CASPT2, and a solvent shell around it is described by a cheaper molecular mechanics methods. For a review of these methods we direct the reader to these recent reviews [27, 28]. Finally, for excited electronic states one must also consider the difference between equilibrium and non-equilibrium solvation; in the latter the solvent system can relax after electronic excitation of the solvent.

2 Review of Previous Studies on the Excited States of Transition Metal Complexes

Having established the background theory involved in excited state computation, and discussed specific challenges that apply to the excited states of transition metal complexes in particular, we now review selected works applying such methods to problems in this field. The number of papers in the literature looking at the excited states, electronic spectra, and photochemistry of transition metal complexes is much smaller compared to those for organic systems. The main reasons for this have already been covered above. Relatively few groups have applied the complex machinery required to decipher inorganic spectroscopy, and even less have looked at the reactive photochemistry. We will review some of the most pertinent papers of these groups from the last fifteen to twenty years or so onwards, when the methods we discussed above started to be used for excited states of transition metal containing systems [29, 30]. This review of previous work is far from exhaustive. We follow this with some more detailed examples taken from our own recent research. We aim to give the general reader an idea of what is possible from modern state-of-the art computational techniques applied to transition metal systems. This is in terms of the types of systems that are possible for study and the type of results one can obtain. We also aim to comment here, and in the next section, on what the current challenges are going forward in this field.

We organize this review of previous works here into some important results of the few groups that have made strong and continued contributions to this challenging field, before focussing on case studies from our own recent work. Within the discussions of each of these the work is presented largely chronologically and by molecule type. Much of the work we will discuss here is on metal carbonyl systems, either binary or substituted complexes. The reasons for this are numerous, and include the fact that such systems are paradigm systems in inorganic chemistry, and serve as models for metal–ligand bonding in more complex ligands. Scheme 1 provide the systems reviewed here together with a number scheme that will be used within the text to reference each molecule as it is discussed.

We begin with the work of Baerends and co-workers who have applied DFT and TD-DFT to a wide range of transition metal systems to study their ground and excited states, and their photochemistry, and have reported much work in this field. One such system studied by this group is $Mn_2(CO)_{10}$ (**3**) [31]. Indeed, binary transition metal carbonyls are perhaps the most studied class of complexes in this area. This is probably because of their ability to act as model complexes for metal–ligand bonding and that they have a lower computational cost compared to other, larger complexes with more 'exotic' ligands. The aim of this study was to better understand the photochemistry resulting from both Mn–Mn and Mn–CO bond dissociation channels. They used early versions of density functional theory called the local density approximation (LDA) exchange potential together with a mixture of double and triple zeta STO basis sets and made a number of conclusions including that the low energy excitations were due to $\sigma \rightarrow \sigma^*$ and $d\pi \rightarrow \sigma^*$

transitions, with ligand field (LF) d → d transitions occurring at higher energy. The metal 4p orbitals were also found to mix in with metal 3d orbitals in these σ and σ* orbitals. Mn–Mn dissociation was thought to occur from the σ → σ* transition and Mn–CO dissociation occurring from the ligand field d → d transitions. These authors revisited the photochemistry of this system again the following year presenting potential energy curves for Mn–Mn and Mn–CO bond dissociation channels. Three different density functional methods were used including the local spin density approximation (LSDA), another early version of density functional theory, with different corrections to the exchange and/or correlation energy. It was concluded that Mn–Mn bond dissociation occurs from the low energy σ → σ* transition. This transition can also lead to axial or equatorial CO loss as the states responsible for these dissociations are near degenerate to the low energy σ → σ* transition at the equilibrium geometry. The energy of this state then decreases below that of the σ → σ* state with the lengthening of the Mn–CO bond. The dominant process upon photoexcitation of (3) above 266 nm was believed to be ejection of a single carbonyl ligand. The authors looked again at the excited states of (3) again in 1999, this time reporting 'the first TD-DFT calculations on the spectra of molecules containing transition metals' [32]. The authors stated that at the time of writing these results were the most high-level theoretical results reported for (3) at that point. They concluded that their results compared well to experimental results. Studies like these illustrate the speed with which advances in theoretical methods have been made and how application to a well-known paradigm system can help drive other applications of the method after calibration with experimental data. With respect to (3) a general understanding of what the chemical character of the experimental bands are in this system has been achieved, but further work is required to get a deeper understanding, specifically with respect to how these excited states relate to the potentially rich photochemistry displayed by this system. This is discussed in the next section.

Aside from (3), the excited states of other binary carbonyls and substituted carbonyls were also studied by Baerends and co-workers in the mid-to-late nineteen nineties. The electronic spectra of group 6 hexacarbonyls $M(CO)_6$ of chromium, molybdenum, and tungsten were studied [33] along with a study of the involvement of ligand field excited states in the M-CO bond dissociation photochemistry of $Cr(CO)_6$ (4) [34]. The main finding of these two studies showed that the lowest energy excited electronic states were not of ligand field character but in fact were of charge transfer character. This assignment disagreed with the accepted view that the lowest excited states were of ligand field character, by which an internal metal d → d excitation is meant. The seminal paper of Beach and Gray from 1963 [35] where they measured and assigned the electronic spectra of the group 6 hexacarbonyls first proposed this by discussing their spectroscopic results in terms of the basic molecular orbitals of these carbonyls. It was the first time such a discussion had been put forward for d^6 octahedral transition metal complexes in an attempt to describe the general features of such isoelectronic complexes in this way. Such topics as the competition between electronic states of very different chemical character (LF vs. MLCT) were covered and the authors correctly assigned most experimental bands.

However they assigned a low energy shoulder on the MLCT band to be of ligand field in character, irradiation of which resulted in CO loss. It is now considered that MLCT states are initially excited and undergo vibronic couplings with LF states to cause ultimate dissociation (see below for discussion of the reactive photochemistry) [36, 37].

Substituted metal carbonyls studied by Baerends et al from this time include the excited states and consequent reactive photochemistry of $Cr(CO)_5CNCN$ (**5**) [38] and $MnCl(CO)_5$ (**6**) [39]. For the case of (**5**) the manner by which the CNCN ligand coordinates to the metal centre was investigated by both experimental and theoretical (TD-DFT) means. It was found that spectroscopic results compared well with experimental (UV–vis, UV-photoelectron and IR) results and further that the ways in which CNCN bonds to the metal centre is similar in some ways to the carbonyl ligand. The σ donor abilities of the two ligands were found to be quite similar while the CNCN ligand is a stronger π acceptor ligand than CO; the authors concluded that the CNCN $3\pi^*$ ligand has a lower amplitude than the $2\pi^*$ orbital on the carbon of the carbonyl ligand, probably due to it being a stronger π acceptor. The photo-dissociation of (**6**) was investigated using DFT to produce potential energy curves in order to probe the initial processes in the dissociative photochemistry. It was found that the initial excited states include the Mn-Cl σ^* orbital at the equilibrium geometry and the potential energy curves of these states are dissociative with regard to both axial and equatorial CO loss. The loss of the Cl ligand does not occur from the lowest excited state which is surprising considering that the initial transition is to its σ^* orbital. This mechanism for the loss of a CO ligand was also found to be the case for other similar complexes, (**3**) and $MnH(CO)_5$ (**7**). However the mechanism by which the σ bond was broken was found to be different for each complex, for example the lowest strongly absorbing state is a $\sigma \rightarrow \sigma^*$ transition that leads to homolytic cleavage in (**3**) and its group 7 mixed metal cousins [40]. Phosphine substituted carbonyls were also studied by the same group, $Cr(CO)_5PH_3$ (**8**) and ax-$Fe(CO)_4PH_3$ (**9**), with respect to their photodissociation mechanisms compared to their binary counterparts (**4**) and (**1**) [41]. They did this with a view to understanding the photodissociation mechanism of larger phosphorous ligands. It was found that both species favour dissociation of their phosphine ligands over their carbonyl ligands with the lowest excited states of (**8**) repulsive with respect to PH_3 but not for the carbonyl ligands, while the lowest excited state of (**9**) was repulsive for both phosphine and carbonyl ligands. Excited state quantum dynamics simulations were performed along the potential energy curve of this state and showed that ejection of the phosphine ligand was heavily favoured over the carbonyl ligand by a branching ratio of 99 to 1.

Another paradigm transition metal system studied by Baerends and co-workers using TD-DFT includes a study on the permanganate anion, MnO_4^- [42]. They looked at this system in order to assign the second and third absorption bands by studying the vibronic structure of the spectrum. The authors stated that the blurred second band in the experimental spectrum was due to a Jahn–Teller active vibration which has a double-well shape of the potential energy surface along this vibrational state of a ^1E state which is formed following a Jahn–Teller distortion of a $^1T_2$ state into $^1B_2$ and the ^1E state. The double-well shape was thought to be due to an avoided crossing with a higher energy state. The third band was found to be much simpler in structure than the second. However both bands exhibit strong configurational mixing meaning that these bands cannot simply be assigned each to a single orbital transition. This is a feature that is now becoming more common to the assignment of the electronic spectra of transition metal complexes. The authors argued that due to good agreement between their results and experiment, that their interpretation of the electronic structure of the permanganate anion is correct. However they also noted that their results for the vertical excitation energies and their spacing did not agree with the experimental spectrum, but the authors argued that environmental effects were not included in their calculations that would have affected the experimental results.

The photophysics of octabutoxy phthalocyaninato-Nickel (II) in toluene was studied in a joint experimental and theoretical study [43]. This particular complex was studied due to its possible use in new cancer therapies, in that it can efficiently absorb photon energy and then rapidly deactivate creating thermal energy that is highly localized that can result in cell death. This study investigated this deactivation process following excitation to the S_1 (π, π^*) state. The authors' DFT results showed that the preferred structure included a D_{2d} saddle-like coordination sphere around the nickel that has an equivalent structure separated from the other by a D_{4h} planar structure, and the structure could readily 'flip' between the two saddle structures. The proposed deactivation mechanism applies to both configurations around the nickel atom. Following excitation to the S_1 state, a transformation occurs to a vibrationally hot $^3(d,d)$ state, which then cools to the lowest vibrational state approximately 8 ps later. There is then a population transfer to a ^{3}LMCT state, with eventual decay to the ground state after 640 ps. For a review summarizing the work of Baerends and co-workers with respect to TD-DFT applied to the excitation energies of different metal complexes and a more thorough discussion of the different TD-DFT methods used, see [44].

Another group that have made significant contributions to theoretically studying the photochemical reactivity and spectroscopy of transition metal systems are the group of Daniel and co-workers. We highlight a number of studies from this group from the start of the last decade that shows what can be achieved using state-of-the-art computational techniques. The lowest excited states of (**3**) were studied using CASSCF and CASPT2 [45]. They concentrated on the lowest energy part of the experimental gas phase UV spectrum, with two main bands centred at 3.31 and 3.68 eV. They assigned these to MLCT and MC transitions, respectively. The authors also noted that at the time of writing that the application of CASPT2 had been 'pushed to its limits' when applied to his problem. Much work has been done by the groups of Daniel and by Baerends on (**3**), applying different DFT and wavefunction-based methods to look at the chemical character, excitation energies and transition moments of this system to test the performance of these methods against a transition metal complex. This case was one of the first transition metal complexes that such methods were applied to, and excited states looked at, gathering some important information. What has been achieved from all these studies on this system is a good general picture of what states contribute to experimental bands in the electronic spectrum, and the chemical character of those states. The exact ordering of these states, and how they interact vibronically with each other still requires some clarification. Nevertheless good qualitative and semi-quantitative information has now been obtained for this species.

The authors also looked at the near-UV–vis electronic spectroscopy of [Ru(X) (Me)(CO)$_2$(Me-DAB)] (X=Cl or I, DAB=1,4-Diaza-1,3-butadiene) (**10**) which serves as a model complex for the larger and more commercially interesting complex [Ru(X)(Me)(CO)$_2$(iPr-DAB)] [46]. These complexes were known for their complex photochemical, photophysical, and electronic properties and were believed to have applications as photosensitizers. It was found that the experimental spectral UV–vis bands could be assigned with CASSCF and CASPT2 however, the accuracy of results was found to be sensitive to the functionals used with TD-DFT, with non-hybrid functionals producing 'meaningless' values. There are many types of charge transfer states in the electronic spectra of this molecule, for example MLCT, halide to ligand (XLCT), and σ bond to ligand (SBLCT).

The excited states of two different W(CO)$_5$ -π-acceptor (**11**) complexes were also studied in another paper using TD-DFT and MS-CASPT2 methods [47]. The two systems studied comprised of W(CO)$_5$ bonded to either 4-[(E)-2-carbomethoxy-2-cyanovinyl]pyridine (**12**) or 4-(2,2-dicyanovinyl)pyridine (**13**). They did this to investigate the character of the lowest excited states compared to those of complexes with weaker acceptor ligands, and to also look into the dynamics of these systems that were investigated experimentally by time-resolved laser flash spectroscopy. They optimized the ground state structures with the B3LYP density functional before using these optimized structures to look at the excited states using TD-DFT, CASSCF, and MS-CASPT2. The authors found that the lowest two excited states of both complexes were of MLCT character from a tungsten 5d orbital to a π* orbital on the ethylenic bridge of the acceptor ligand. It was believed that little photoreactivity would occur following irradiation to these MLCT states. Irradiation of higher lying LF states of these complexes might result in ligand substitution of these complexes, but the possibility of reaching these states thermally in significant quantum yields was not believed to be possible. This theoretical work therefore helped to confirm the experimentally observed photostability in such complexes following low energy photoexcitation. CASSCF and CASPT2 methods were applied to study of the emission spectroscopy of the MLCT states in HRe(CO)$_3$(H-dab) (H-dab=1,4-diaza-1,3-butadiene) (**14**) [48]. The authors described this as a model complex for a range of different α-diimine transition metal carbonyls. CASSCF and MRCI methods were used to produce potential energy curves of the Re-H bond elongation for the ground state, low lying MLCT states and a ^{3}SBLCT state that was also believed to play a meaningful part in the photochemistry of this system through a ^{1}MLCT to ^{3}SBLCT intersystem crossing leading to homolysis of the Re-H bond. This process was thought to be in competition with emission from the ^{1}MLCT state. No other degrees of freedom were investigated. Wavepacket propagation dynamics were then performed to produce the emission spectra in this 1D case. The theoretical results managed to correctly generate the main features of the spectrum that represent these two photo-activated processes but the authors stated higher level theoretical investigations would need to be done to explore the competition between them.

Spin-orbit effects were investigated on the electronic spectroscopy of two other metal carbonyl complexes around the same time. This time the complexes were dihydride species $H_2M(CO)_4$ (M=Fe, Os) (**15**) [49]. They investigated both singlet and triplet low lying excited states of both complexes using a two-step approach. The first step used state-averaged CASSCF and CASPT2 calculations to calculate the spin-free energies of the first eight electronic excited states of singlet and triplet spin of A_1, B_1, and B_2 sub-groups of the C_{2v} symmetry complexes. In the second step the authors considered spin–orbit interactions between the various states by calculating the spin–orbit integrals using an effective Hamiltonian sourced from the uncontracted spin–orbit configuration interaction (SO-CI) method. Results from these two steps were then combined. It was found that spin–orbit effects had little influence on the absorption spectrum of the iron complex but did have a considerable effect on the spectrum of the osmium complex. This included a large mixing between the lowest lying states and a large splitting for the lowest triplet B_2 state. MS-CASPT2, CASSCF, and TD-DFT methods were also used to look at the spectrum of two cobalt-based complexes $RCo(CO)_4$ (R=H,CH$_3$) (**16**) [50]. They did this to further study the electronic spectra of these hydrides using state-of-the-art methods and also to compare their results for this complex with the methyl complex which is a model complex for femtosecond laser control spectroscopy. It was found both complexes absorbed from around 4.33 eV and both have a strong band centred at 4.95 eV. It was believed that SBLCT and MLCT states contributed to these bands in agreement with experimental assignment. It was also concluded due to the similarities in their spectra that the photoreactivity of these complexes would also be quite similar. The role of MLCT states was investigated in the *trans–cis* photoisomerization of the styrylpyridine ligand in [Re(CO)$_3$(2,2′-bipyridine)(*t*-4-styrylpyridine)]$^+$ (**17**) in a paper by the Daniel group from 2006 [51]. DFT methods were used to optimize the ground state structures of possible *trans* and *cis* isomers, with a barrier between *trans* and *cis* isomers of 27.0 kcal mol^{-1}. TD-DFT, CASSCF, and CASPT2 were then used to calculate the electronic spectra of these isomers. The lowest excited states of the complex were of MLCT character with a singlet intraligand state nearby. One-dimensional potential energy curves were produced with the CASSCF method as a function of the C=C bond torsion angle on the styrylpyridine ligand. It was found that the states that are important for photoisomerization of the styrylpyridine ligand are different between the free ligand and when it is in the complex. In the free ligand the singlet intraligand state determines the mechanism, in the complex this happens through a triplet intraligand state. The low-lying MLCT states are important in this mechanism and the metal was thought to act as a photosensitizer in this case. The same methodology was used again to study the photochemistry of $CH_3Mn(CO)_5$ (**18**) in which the authors endeavoured to identify the electronic excited states involved in experimentally (in solvent) observed results in which an equatorial carbonyl ligand is lost preferentially to the methyl ligand following excitation above 4.02 eV. The excited states were of MLCT (metal 3d to ligand π^*) character and a large density of these states was present between 4.69 and 6.01 eV. The authors reported that the process is a very complicated one, a common feature with transition metal complexes, and that several minor channels for dissociation such as loss of an axial CO ligand, are present. Analogues of (**16**) were also studied by this group [52] in a similar way but this time with both singlet and triplet states, but also using time-dependent wavepacket propagations as a function of both Co-CO (axial) and Co-R bond cleavages. Two main peaks were found in the absorption spectrum of each complex and assigned to MLCT (metal 3d to ligand π^*) states. Photoactive metal to σ bond charge transfer (MSBCT) and MC states were identified for both complexes that result in cleavage of either Co-CO (axial) or Co-R bonds. Binary carbonyls have also been studied by Daniel and co-workers, including (**4**) to which a number of different methods were applied [53, 54]. They did this in order to try and get theoretical results that compared well with experiment, so being able to assign the character for each state. They also wished to "illustrate the theoretical difficulties inherent" to binary transition metal carbonyls. As with previous studies on other metal complexes MS-CASPT2 was used along with CASSCF and TD-DFT, however only MS-CASPT2 gave results that were "completely quantitatively satisfactory" provided the active space was large enough, with TD-DFT results being sensitive to the choice of functional. Indeed, the authors labelled (**4**) as a "tough case".

The application of equation-of-motion CCSD (EOM-CCSD), a method based on the correlated coupled cluster singles and doubles (CCSD) developed for obtaining external states to the CCSD reference, e.g., ionized, attached, or excited states [55, 56], with a range of all-electron basis sets including atomic natural orbital basis (ANO) sets yielded accurate results for this system. Both singlet and triplet states were studied in order to get an idea of the initial photochemistry of this system and the important states involved. The results of the EOM-CCSD calculations showed the presence of a weak shoulder at 3.90 eV that was not produced by the results of their previous paper. The two main spectral bands were also found with positions close to experiment at 4.37 and 5.20 eV and both assigned as MLCT states. As with previous studies one-dimensional potential energy curves were calculated, for (**4**) as a function of Cr-C bond length and the authors found that the lowest LF state at 4.37 eV is dissociative for CO loss and also near degenerate with the lowest absorbing state, with an avoided crossing between the two at the equilibrium bond length. The authors argued that it was these features with the lowest electronic states that explained experimentally observed CO loss in under 100 fs [57]. Furthermore this type of photodissociation mechanism appears to be applicable to a number of binary transition metal carbonyls, that also include Jahn–Teller-induced distortions in their vibrationally hot photoproducts [57–61] (see below for further studies on this aspect). This group have also applied the EOM-CCSD method to other smaller transition metal complexes such as the $NiCH_2^+$ cation [62], where it could be applied without any issues. The spectroscopy and excited state dynamics on the rhenium-based complex $(CH_3)ReO_3$ (**19**) has also been studied at length.

They used a similar approach to their previous papers, plus 1D wavepacket propagation studies were also carried out for the lowest singlet and triplet states. The authors found that TD-DFT results with the PW91 functional based on DFT optimized structures gave excellent agreement with experimental spectra with CASSCF and MS-CASPT2 unable to improve on this. The lowest energy experimentally observed transition was assigned as an LMCT state from an oxygen p orbital to a d orbital on the rhenium atom. The second and third transitions were also assigned as LMCT states. Results with different density functionals gave little difference to their best results. The results of wavepacket propagations on the low-lying singlet and triplet states, with special attention paid to the role of spin–orbit coupling in the photodissociation mechanism, found that the second lowest state was the initial absorbing state that was of $^1A_1$ symmetry. This state is coupled non-adiabatically to the lowest bright state of ^1E symmetry with the state responsible for homolysis of the Re-CH$_3$ bond, $^3A_1$, close in energy. Following population of the initial absorbing state the two latter states were found to be populated within a few tens of femtoseconds, with an intersystem crossing between these, controlling the photodissociation. Recent studies look at the photoisomerization mechanisms of rhenium(I) complexes with ligands larger than the ones discussed above e.g., diimine complexes [63–66]. It is an important point in these papers that spin–orbit coupling is a large factor in the photoisomerization mechanism of such complexes and that interplay between singlet and triplet states allows this chemistry to proceed. Larger copper(I)-based phenanthroline complexes have also been studied [67]. It is only through theoretical study that the coupled potential energy surfaces can be probed and the chemical character of transitions in the electronic spectra accurately determined.

Finally in this section we discuss some of the results of Ziegler and co-workers and they have done much work in this field. This includes developing a method for accurately studying the UV–vis spectra of molecules with spatially degenerate ground states, applied in one paper to d^1 and d^2 transition metal complexes with either tetrahedral or octahedral symmetry [68]. This method is called 'transforms reference via an intermediate configuration Kohn–Sham TD-DFT' (TRICKS-TDDFT). It was designed to remove the difficulties when one has to deal with a multi-reference ground state and how this can affect the results for electronic excited states calculated from that ground state. This was done by replacing the ground state with that for a nondegenerate excited state with certain desirable properties depending on the system. It was found that this method produced results that mostly agreed with experiment for the values of transition energies of the complexes studied. One issue discovered by the authors was that in some cases the calculated transition densities were to states that were not spin eigenfunctions. The authors suggested a method to counter this based on the SCF sum method, which helped to correct this weakness. An example of the types of system reported by Ziegler and co-workers is a TD-DFT study of the platinum (II) complexes [PtCl$_4$]$^{2-}$ (**20**), [PtBr$_4$]$^{2-}$ (**21**) and [Pt(CN)$_4$]$^{2-}$ (**22**), all of which are square planar [69]. This was done to test the performance of their two-component relativistic TD-DFT method that accounts for relativistic effects in both scalar and spin–orbit coupling forms. The systems studied are all Pt $5d^8$ closed-shell and the transitions studied came out of the 5d orbitals of the platinum. The authors noted that these transitions are especially susceptible to spin–orbit coupling effects. Very good agreement between their method and experiment was found, with a caveat that a suitable exchange correlation potential is used. Another paper used TD-DFT to look at the LF states of $[Co(en)_3]^3$ and $[Rh(en)_3]^{3+}$ and the lowest energy states of $[Fe(phen)_3]^{2+}$ (**23**) (en=ethylenediamine and phen $= 1,10$-phenanthroline) [70]. The authors studied the circular dichroism of these states and focused on the impact of charge transfer corrections in 3d transition metal complexes. A range of different density functionals was tried on each complex and it was noted that the excitation energies are very sensitive to the choice of functional, whereas for spectral intensities the opposite is the case. The authors believed that meaningful information could be extracted from assigning the CD spectra of these complexes. It was also found that the iron complex studied here had a dense spectrum that made it difficult to assess the accuracy of the theoretical spectrum due to a complicated interplay between the ability of a functional to accurately optimize the ground state geometry of the complex, and the different types of excitations present in the spectrum.

This list of contributions to this field in recent years is far from exhaustive, for a more in-depth discussion we direct the reader to the following reviews [30, 44, 71]. We now present in more detail aspects of the ongoing work in our group in this field. This is in order to further illustrate to the reader the type of chemistry and photo-induced effects that can be investigated using state-of-the-art computational methods on transition metal complexes and also complement the type of work already discussed above. The work presented below covers some of the salient aspects in theoretical study of the photochemistry of transition metal complexes, namely photodissociation, photoisomerization, non-adiabatic relaxation, Jahn–Teller and pseudo-Jahn–Teller effects, and the accurate analysis of electronic spectra of various transition metal complexes.

The first class of complexes we discuss are binary transition metal carbonyls and our ongoing work regarding their photodissociation mechanisms. These are paradigm complexes in inorganic photochemistry. In recent years ultrafast pump-probe laser-based experiments [57–61] have shown that these carbonyls possess a rich, complex photochemistry involving vibronic coupling-related phenomena that is still not fully understood (see above for discussion of some other studies applied to this important class of molecules). Modern multiconfigurational methods such as CASSCF and CASPT2 applied to these systems have provided information on the types of processes that are involved following electronic excitation of these carbonyls using an ultrafast laser pulse. Broadly speaking individual study of different carbonyls such as (**4**), (**3**), and (**1**) with both experimental and theoretical methods have shown common features in their photodissociation mechanisms. Following excitation with a pump laser pulse into a dense manifold of charge transfer states an initial photoproduct is formed on the order of tens to hundreds of femtoseconds. The initial photoproduct in all cases is believed to be the parent complex minus one carbonyl ligand or in the case of (**3**) either one carbonyl ligand is ejected or Mn–Mn bond homolysis occurs. This photoproduct can then relax to the lowest state within its respective spin manifold also on the timescale of femtoseconds and then persevere in this state for many picoseconds before going on to dissociate further. Vibrational coherences observed in the transient absorption spectra of these species indicate that these initial vibrationally hot photoproducts have all their excess energy in very specific vibrational mode(s), analysis of these modes identifies them as CO-M-CO bending vibrations. This raises the question of how the coherent vibrationally hot photoproduct is formed in such a timescale? Theoretical results allow a mechanistic picture to develop. The initial step in this process involves excitation of the ground state complex into a manifold of charge transfer states (the nature of the initial excited state in these complexes is discussed further below). Following this, the system may switch between the states in this manifold several times before eventually transferring to a dissociative ligand field state. This is due to very strong vibronic couplings among the degenerate bright CT states, and also the dark CT and LF states. This ultimate LF state is dissociative along an M-CO bond resulting in ejection of a single carbonyl ligand within one or two vibrational periods of an asymmetric M-CO stretch. For the bimetallic (**3**) this is also the case in the CO dissociation channel although the dominant channel involved homolysis of the Mn–Mn bond via direct population of the dissociative $\sigma \rightarrow \sigma^*$ state. The remaining ligands then begin to fill the resulting coordination hole around the metal. This results in a highly symmetric geometry, within the same spin manifold, that is Jahn–Teller active, in an orbitally degenerate electronic state. This geometry is a point on a conical intersection seam between the two components of the degenerate state. The system can then pass through the intersection from the upper excited state to the lower electronic state by undergoing the Jahn–Teller distortion, which are the standard branching space motions of a general conical intersection. The system can then persevere on this lower surface undergoing motion in the moat around the conical intersection (classically thought of a pseudo-rotation). The barriers to pseudo-rotation on the lower surface of the intersection are formed by higher-order pseudo-Jahn–Teller type couplings between the (previously) degenerate components, or the ground and higher excited states, and can generate a complex surface topology in this region (see, for example, Fig. 3 below for $Cr(CO)_5$ and $Mn(CO)_5$ photoproducts). We loosely aim to use a three-phase approach in order to study such photochemistry in these complexes, namely spectroscopy, potential energy surface modelling, and dynamics, which will now be explained in more detail. The first phase is to accurately compute the electronic spectrum and within the appropriate spectral energy range assign all coupled vibronic states. This in itself produces a number of challenges, and we shall discuss this along with other transition metal complexes later in this review.

The second phase is to map the topology of the excited state potential energy in regions most important to the photodissociation process. This of course includes regions of strong non-adiabatic coupling, for example the point of conical intersection connecting excited states to the ground electronic state. Multiconfigurational methodologies like CASSCF and CASPT2 are needed in this case. As discussed earlier when using CASSCF a suitable active space is needed and a great deal of chemical intuition is needed when deciding how many electrons and the correct

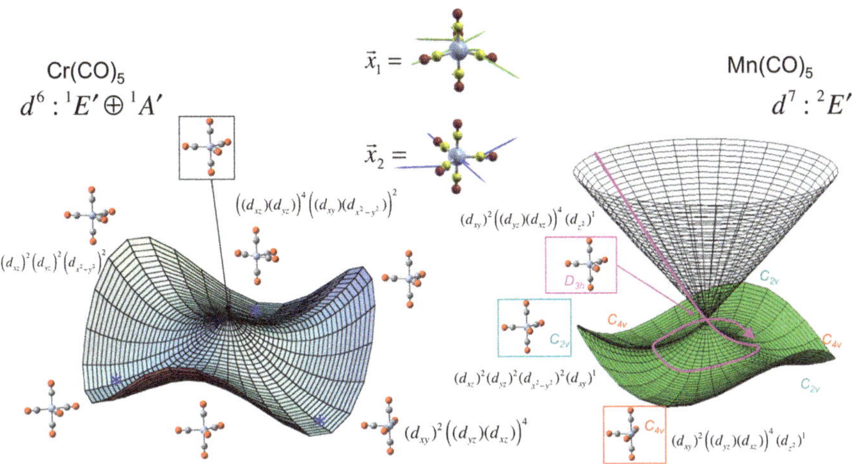

Fig. 3 Surface topology around Jahn–Teller conical intersections for Cr(CO)$_5$ and Mn(CO)$_5$ photoproducts formed by photodissociation of Cr(CO)$_6$ and Mn$_2$(CO)$_{10}$

orbitals to include for the system and chemistry under study. The 3d orbitals would seen an obvious first choice in this case, so making a (6,5) active space (6 electrons in 5 active orbitals) for the Cr(CO)$_5$, (7,5) for doublet (7 electrons in 5 active orbitals) Mn(CO)$_5$, and (8,5) for Fe(CO)$_4$ unsaturated photoproducts (8 electrons in 5 active orbitals). It was discovered, however, that this active space was difficult to converge, and the wavefunctions that resulted sometimes suffered from spurious symmetry breaking. Such an active space was greatly improved by effectively doubling the active space by including an equivalent 4d counterpart of each 3d orbital already in the active space, but with an extra node in the internuclear M–L bonding region. This effectively allows for the effect of dative bonding in an M–CO bond, a dominant feature of this type of bond, since it variationally introduces dynamic electron correlation into the wavefunction allowing the donor pair electrons to be on average further apart. This is shown in Fig. 4 below.

This results in a (6,10), (7,10), and (8,10) active space for the three systems mentioned above. For a 3d orbital that chemically would be considered doubly occupied a (6,5) active space has an occupation of around 1.99, while a (6,10) active space has the 3d orbital occupation around 1.8–1.9, and the 4d orbital occupation around 0.1–0.2. A variant of this type of doubling has been successfully implemented before for calculating metal carbonyl bonding in the ground state using CASPT2 [72].

CASSCF calculations have found Jahn–Teller geometries present in the photodissociation pathways for these systems, with tetrahedral symmetry for singlet Fe (CO)$_4$, and a D$_{3h}$ trigonal bipyramid for singlet Cr(CO)$_5$ and doublet Mn(CO)$_5$. The topology on the lower surface around the point of intersection has a number of symmetry equivalent minima separated by equivalent transition states (see Fig. 3). In the case of Cr(CO)$_5$ and Mn(CO)$_5$ the surfaces are very similar involving three

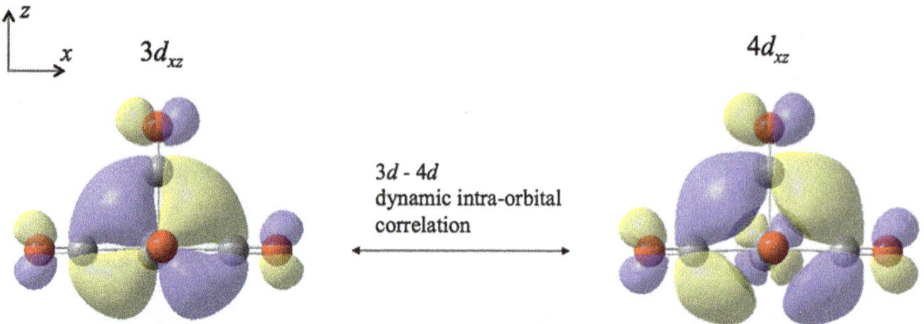

Fig. 4 Orbital doubling in CASSCF wavefunctions to better describe dative boding between ligand and metal. (reproduced from reference [37])

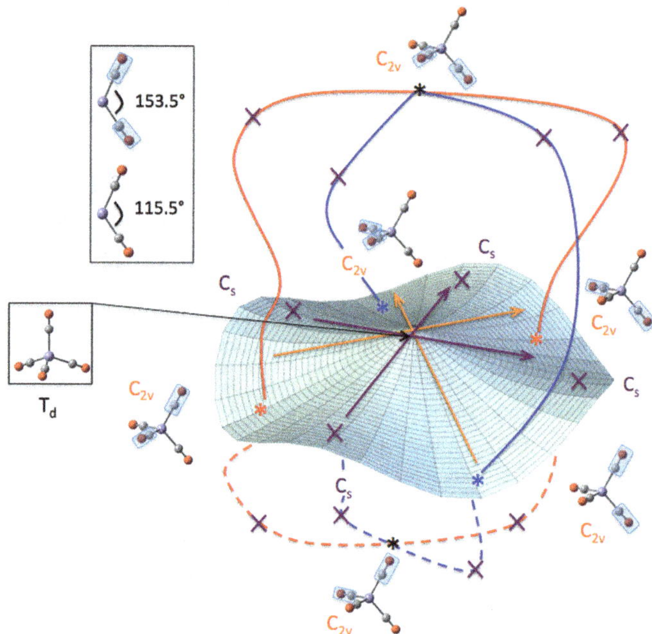

Fig. 5 Surface topology around triply degenerate Jahn–Teller conical intersections for Fe(CO)$_4$. Colours indicate pathways connecting minima and transition states. (reproduced from reference [37])

C$_{4v}$ square pyramidal minima separated by three C$_{2v}$ transition states. For Fe(CO)$_4$ this has six equivalent C$_{2v}$ symmetry minima separated by 12 equiv. C$_s$ symmetry transition states (shown schematically in Fig. 5).

The barrier heights for Mn(CO)$_5$ and Fe(CO)$_4$ are less than 2 kcal mol^{-1} indicating effectively free pseudo-rotation in the vibrational hot photoproducts. For Cr(CO)$_5$ there is a more substantial barrier ~12 kcal mol^{-1}, and dynamics simulations (see below) indicate that the pseudo-rotation may become trapped around a single minimum. This is consistent with experimental observation of coherent oscillations with a frequency of a magnitude that matches an OC-M-CO bending vibration (which is a linear combination of the branching modes). Further information on each surface and how they were calculated can be found elsewhere [36, 37]. An important point to note is that this type of behaviour appears common to binary carbonyls and we believe that such a photodissociation scheme should be applicable to most binary carbonyls (with an unfilled d-shell so excluding $Ni(CO)_4$ (**24**)). It will be shown in the following examples that very similar non-adiabatic features in the photochemistry of general transition metal complexes are the norm rather than the exception.

The third and final phase is to run dynamics simulations on these coupled potential energy surfaces. There are a number of different types of methods available to do this each with an associated computational expense and various advantages and disadvantages. Considerations include how to calculate the potential energy surfaces. For example, model Hamiltonians (e.g., vibronic coupling Hamiltonians) may be constructed and fitted to multidimensional functions for all or some of the vibrational degrees of freedom of the system. Alternatively potential surfaces may be computed locally using on-the-fly electronic structure energies, gradients, Hessians, and so on. Secondly one must consider how to describe the nuclear motion on these potential energy surfaces, which can be treated as semi-classical particles with well-defined trajectories or via quantum wave packets. Such dynamics simulations will not be discussed further here but see reference [36] for a review, contrasting the various approaches for $Cr(CO)_5$ relaxation.

It is also apparent that non-adiabatic interactions between ground and excited states in transition metal systems can affect not only their photodissociation mechanisms but also their ground state structure. This was the case in a study from 2009 in which the structure of a bimetallic molybdenum complex was not what would be expected on the basis of simple electronic structure arguments [73]. Cotton and co-workers [74] found that the edge-sharing bioctahedral complex $Mo_2(DXyIF)_2(O_2CCH_3)_2(\mu_2\text{-}O)_2$ (DXyIF=N,N'-di(2,6-xylylformamidine) (**25**) had a central $Mo_2(\mu_2\text{-}O)_2$ ring that was not a D_{2h} symmetry square shape as one would expect but rather had a distorted C_{2h} symmetry rhomboidal shape. The authors hypothesized that this could be due to a pseudo-Jahn–Teller effect (two or more non-degenerate states coupling via a non-degenerate vibration of appropriate symmetry) and not crystal packing distortions.

CASSCF calculations were performed on a simplified model of this system to investigate this based on a diagnostic test utilizing symmetry restricted analytical CASSCF Hessians [75]. The CASSCF active space used was a (4,6) active space including σ, π, and δ orbitals and their antibonding equivalents. DFT calculations were also performed using the B3LYP functional on both model and target geometries. DFT was used to see in part if the ground state structure did indeed include a rhomboidal central motif, and whether it agreed with CASSCF predictions where applicable. Both DFT and CASSCF also agreed that when the central motif was constrained to a square D_{2h} symmetry the geometry was a first order saddle point (transition state). The imaginary vibrational mode was found to be of B_{1g} symmetry

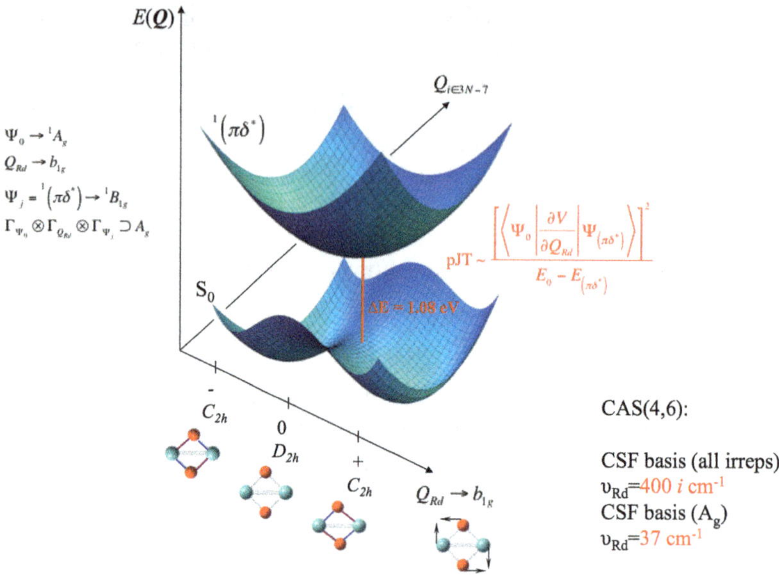

Fig. 6 Ground and excited potential energy surfaces along the rhomboidal distortion in $Mo_2(DXyIF)_2(O_2CCH_3)_2(\mu_2\text{-}O)_2$ as calculated by CASSCF. A pseudo-Jahn–Teller mixing of ground and excited $^1(\pi\delta^*)$ states causes the molecule to adopt a lower symmetry structure. (reproduced from reference [73])

(within the D_{2h} point group symmetry at this point on the surface) and connected to two equivalent rhomboidal minima. Through the CASSCF hessian diagnostic test this vibration was found to vibronically couple the ground state and the first excited state S_1, which has B_{1g} symmetry and was of $^1(\pi\delta^*)$ character, and has its minimum at D_{2h}. The energy gap between these states was found to be 1.05 eV at the D_{2h} structure (Fig. 6). The CASSCF pseudo-Jahn–Teller diagnostic proceeds by calculating the Hessian on a basis of symmetry restricted configuration state functions (CSFs) and compares with the same calculation in the basis of all CSFs. Note that the energy and gradient (first derivative) only requires CSFs of a single symmetry and does not couple those of different symmetry. The Hessian, on the other hand, contains derivative coupling like terms that couple CSFs of different symmetry. Thus the Hessian eigenvectors evaluated in each of these bases will have different eigenvalues. For the pseudo-Jahn–Teller effect to be operative one notices a change from positive to negative curvature as the full CSF basis is used. For this system using only CSFs of A_g symmetry gives positive curvature to the rhomboidal distortion, while this becomes negative when those of B_{1g} are included. The dominant configuration for the first state of B_{1g} is $^1(\pi\delta^*)$. Thus in this system, at D_{2h} geometries, although the ground and first excited $^1(\pi\delta^*)$ state are separated by around 1 eV, the vibronic coupling between these states is strong enough to overcome this energy difference (i.e., the mixing of states along the rhomboidal distortion is larger than the energy difference to give an overall lowering of energy) and causes the ground state minimum geometry to be C_{2h} rhomboidal (where the adiabatic electronic state now contains a small admixture of $(\pi\delta^*)$ configuration in addition to the dominant closed shell $\sigma^2\pi^2\delta^2$ configuration). This is summarized in Fig. 6 below.

Vibronic coupling between ground and excited states can also affect the way in which systems can isomerize when electronically excited. This was the case in a report on the photoisomerization mechanism of a platinum containing amido pincer complex [76]. The complex studied was $(BQA)PtMe_2I$ (BQA=bis(8-quinolinyl) amine (**26**) and was studied after some interesting results from Harkins and Peters [77] in which after this complex when irradiated with visible light underwent an unexpected *mer* to *fac* isomerization (all coordinating nitrogens and metal being coplanar to all coordinating nitrogens *cis* to each other). Both isomers were found to have similar stability from DFT calculations. TD-DFT calculations on the electronic absorption spectrum provided good agreement with the experimental spectrum in terms of observed bands. The important conclusion to be drawn from these results was that the low energy absorption band was predominantly of $\pi\pi^*$ character with little involvement from the metal. It was previously believed that the transitions were of LMCT character. Furthermore the population of a π^* orbital caused delocalization over the amide linker on the BQA ligand. This is important in explaining the photoisomerization as this allows the BQA ligand to fold in. TD-DFT and CASSCF geometry optimizations also found an S_1 *fac* geometry minimum. CASSCF calculations with a (14,12) active space (based on quasi-natural orbitals) located a non-adiabatic radiationless decay pathway via this *fac* geometry. It is a sloped conical intersection connecting the excited to the ground state and sits 10 kcal mol^{-1} above the optimized *fac* minimum. This photochemical feature was also found to be a feature of the bare BQA ligand showing the metal acts as a scaffold in this case. Tying all this information together the photochemical pathway to explain the experimental observations was concluded to start from an excitation localized on the BQA ligand of $\pi\pi^*$ type which delocalizes electron density over the amide link. This provides a driving force towards the *fac* isomer as the BQA can then fold. This then forms an excited *fac* minimum that is connected to the *fac* ground state minimum through a sloped conical intersection; the sloped feature of the intersection provides an additional driving force to produce the ground state *fac* minimum. Once on the ground state the *fac* photoproduct is stable. The metal participates as (or provides) a scaffold that pins the BQA ligand in place in both *mer* and *fac* geometries without becoming directly involved in the photochemistry itself. It was also found that there was no direct ground state pathway for this process and that this non-adiabatic process occurs only on the *fac* side, and is thus photochemically irreversible as *fac* excitation results in radiationless decay back to the *fac* ground state minimum. A schematic of this process is shown in Fig. 7 below.

We have also found non-adiabatic features in the photochemistry of more complex metal complexes such as the photoaquation chemistry of (**2**) [78], with the chromium in an octahedral coordination sphere and the three bidentate tn ligands around it. The mechanism of the photoinduced incorporation of a water molecule to the metal centre was found to share similarities with the simpler

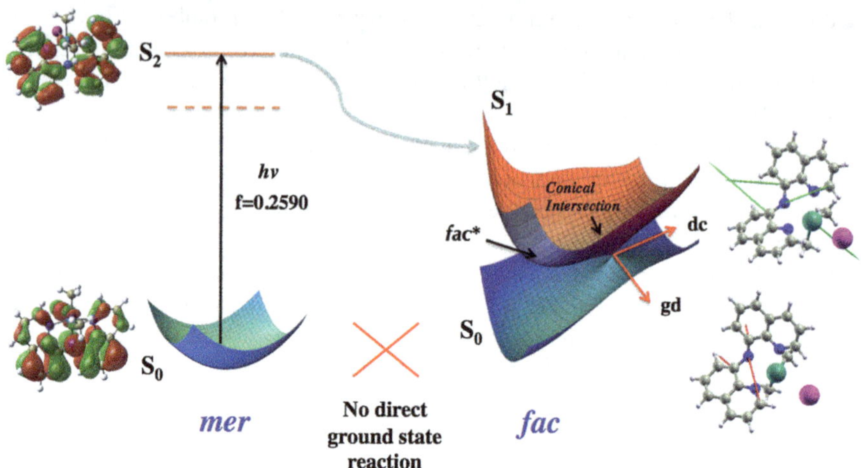

Fig. 7 Schematic of *mer-fac* photoizomerization in (BQA)PtMe$_2$I involving localized $\pi\pi^*$ excitation on the BQA ligand, followed by relaxation to S$_1$ *fac* minimum, followed by radiationless deactivation through a sloped conical intersection connecting the S$_0$ and S$_1$ states. (reproduced from reference [76])

carbonyl systems discussed above. Here the initial photoexcitation causes one end of a tn ligand to dissociate, followed then by a conical intersection (this time between quartet states) connecting the excited and ground electronic states. Different photoproducts can be formed in this process, as shall be discussed. As with the previous studies discussed above, DFT, TD-DFT, and CASSCF methods were applied to this system. The CASSCF active space used was based on quasi-natural orbitals with 9 electrons in 10 orbitals. B3LYP was used to optimize the ground state structures of the initial complex and various photoproducts formed after the cleavage of a single Cr–N bond and the incorporation of a water molecule. TD-B3LYP calculations were also used to map out the electronic excited states of these photoproducts. The overall photo-induced process is rather complex, and in this case the involvement of excited state manifolds of different spin and their interaction can have a crucial influence in the resulting photochemistry. Initial excitation from the ^{4}A ground state results in population of a ^{4}T state. From here the molecule can then undergo intersystem crossing to a ^{2}E state that can then quench the system and halt any further reaction. The main photochemistry results directly from the ^{4}T state however. Following population of this state a Cr–N bond from one of the ligands is cleaved and the ligand can then 'unfold' and the ligand 'arm' can then swing away leaving a five coordinate Chromium that has a Jahn–Teller like geometry. Direct reconnection of the ligand in the ground state is then improbable due to the distance created between the metal centre and the leaving nitrogen. In a similar way to the binary carbonyls Cr(CO)$_5$ and Mn(CO)$_5$ above the remaining connected ligands then move to fill the coordination hole and the coordination sphere around the metal becomes trigonal bipyramidal (Jahn–Teller

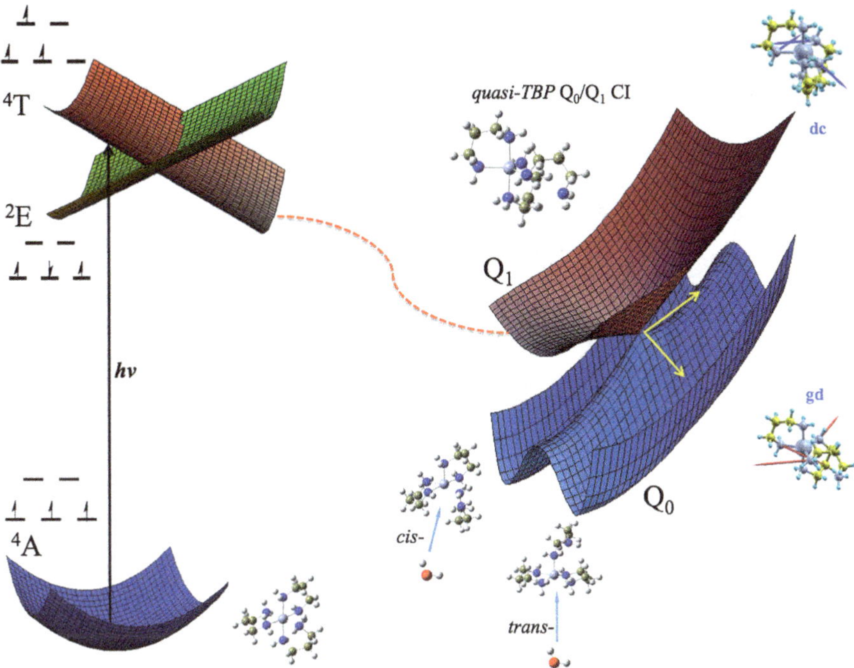

Fig. 8 Schematic of $[Cr(tn)_3]^{3+}$ potential energy surfaces showing conical intersection connecting ground and excited quartet states. (reproduced from reference [78])

like). This geometry is a point of conical intersection that is sloped and allows the system pass from the Q_1 to Q_0 state. Note that the symmetry is not exact here and hence the surface topology is less symmetrical and sloped rather than peaked here. A square pyramidal coordination sphere is formed around the chromium after passing through the intersection, again similar to the binary carbonyls discussed earlier. A water molecule can then coordinate to the metal centre. The surface topography on the Q_0 state around the point of intersection is divided into two channels separated by a barrier relating to the position of the vacant site on the metal centre relative to the partially dissociated ligand. This is classed here as either being *cis* or *trans* to the 'unfolded' ligand. Both of these mono-aqua isomers were found to be isoenergetic. A pictorial representation of this discussion is shown in Fig. 8 below.

This again illustrates how excited states in metal systems can lead to new products and can also influence the types of product formed. Another interesting point to note is that the non-adiabatic pathway is similar between this system and the binary carbonyls in terms of the coordination sphere of the metal. This lends credence to our belief that the binary carbonyls can act as model complexes in terms of photo-induced chemistry.

We have also studied similarly complex cases involving multiply coupled potential energy surfaces and the influence of different spin manifolds in the

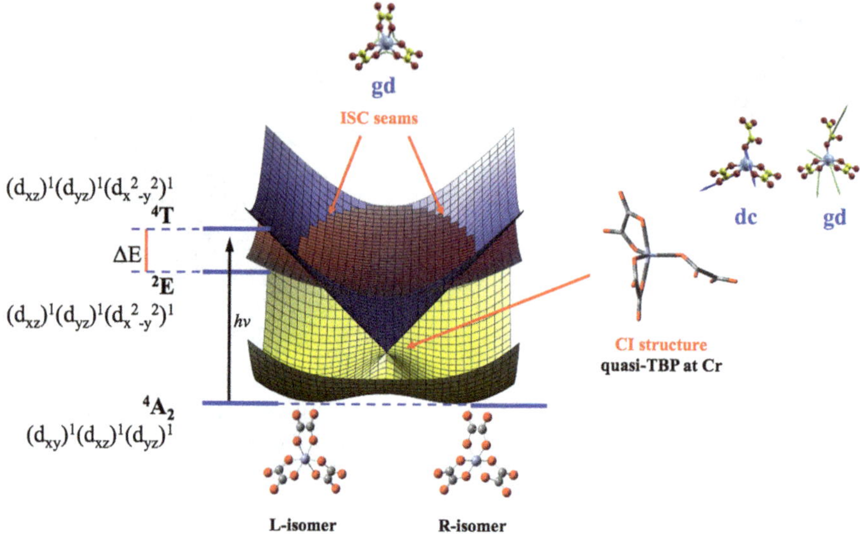

Fig. 9 Potential energy surfaces of $[Cr(C_2O_4)_3]^{3-}$ complex showing conical intersection connecting ground state optical isomers with photoexcited states. (reproduced from reference [79])

relaxation pathways of other systems such as chromium oxalate, $[Cr(C_2O_4)_3]^{3-}$ (Fig. 9) [79], which involves quartet and doublet spin manifolds, conical intersections and seams of intersystem crossing and how they influence each other. The Cr (III) centre here undergoes similar photochemistry to that discussed above for (**2**). The crucial difference here is that the dissociated Cr–O bond can re-connect on the ground state after the five-coordinate degenerate species is reached. Here the five-coordinate trigonal bipyramid structure is again a conical intersection between the ground and first excited quartet states. Motion away from this conical intersection generates either the R- or L- optical isomers of the system. Thus this system is an example of one capable of undergoing photoracemization. The schematic potentials are shown in Fig. 9. For more details see [79].

We finish by briefly discussing our experiences of applying coupled cluster response theory to transition metal complexes. We have applied the CC hierarchy discussed above to first row binary metal carbonyls, [37] and small metal oxides such as TiO_2 nanoclusters [11], and $MnO_4{}^-$. The first thing to note is that the basis set requirements of such application are much more severe than when applying TD-DFT. Both correlated all electron and effective core potential bases have been investigated. For some states an oscillatory convergence in excitation energy is clear, e.g., for the first bright T_{1u} state of (**4**) we see CCS $= 5.553$ eV, CC2 $= 4.150$ eV, CCSD $= 4.990$ eV, CC3 $= 4.374$ eV. The CC3 result is in good agreement with the experimental band maxima. Notice how the excitation energy oscillates to convergence. For some systems though the low cost CC methods such as CC2 break down (see, for example, [11]). This was surprising at first as this tends to be a very robust method in organic systems. It appears that even when the ground state is quite well represented by a single configuration, the static electron correlation in several transition metal complexes is significant enough to cause problems in the response treatment using the CC methods that are only correct to second-order. An example in metal carbonyls is the first $^1T_2$ state of (**24**) where the CC2 excitation energy is 1.361 eV compared with 4.998 eV for CCSD. While single reference methods work well in describing the ground state of (**24**) the doubly excited $3d^8 4s^2$ configuration seems to be important enough that lower cost response methods that fail to treat this with the $3d^{10}$ configuration in a balanced manner have problems describing certain excited states. One can also think that if the single reference response excited states (CCS) that the higher orders correct are often such bad representations of the electronic states that lower order (and cost) methods cannot correct them. For example, CCS excitation energies are usually out by at least 1 eV and often several. The situation for metal oxides is even more severe with some perturbation-based excited state methods completely breaking down [11]. It thus seems to be a general conclusion that CCS and CC2 are not appropriate to study excited states of general transition metal complexes, but that CCSD response provides a very good starting point. This has also been seen in the EOM-CCSD application to (**4**) by Daniel and co-workers discussed above. EOM-CCSD is an alterative formulation of LR-CCSD that generates exactly the same excitation energies (but slightly different transition properties). The effect of triples is often quite small but can affect excitation energies by up to an eV for difficult states. Finally we note that LR-CCSD was able to correctly assign both the OPA and TPA in (**1**), a system that has been studied for its non-linear TPA-induced photochemistry. This brief discussion barely scratches the surface of the application of correlated wavefunction response theories in this field although we hope it gives the reader an idea of the extra complexity involved over application to an organic chromophore. In upcoming publications we will discuss these issues for specific systems and present benchmark results against which other excited state methodologies can be compared.

3 Conclusions and Future Challenges

We have illustrated in this review some of the issues that arise when studying the excited states and associated photochemistry of transition metal complexes using theoretical and computational methods. The complexity of their photochemistry means that many photoreactions are just beginning to be understood. With the advancement of both theoretical methods, laser-based experimental techniques and computational power, photoinduced reactive pathways in transition metal complexes can now be accurately studied. Significant challenges to the theoretician remain though, primarily the size of system that can be studied by ab initio means, the high density of non-adiabatically coupled states, correct simulation of dynamics on coupled potential energy surfaces, environmental effects, and so on. Continued advancement in hardware means that the methods discussed above will be more routinely applied to excited state problems in small and medium inorganic and organometallic systems as it becomes quicker and easier to run such calculations, and continual development of TD-DFT methods makes them more trustworthy for complex transition metal systems, particularly with regard to charge-transfer bands. However, a barrier to general application exists that will require significant methodological and algorithmical advancement. Martinez and co-workers have shown that Kohn–Sham TD-DFT at present is severely challenged, and indeed breaks down, by conical intersections and the so-called doubly excited states [80]. Clearly fruitful areas of study are the extension of DFT to cover multi-reference states and non-adiabatic couplings. It is our belief that combining the cost and ease of use of DFT, with the power and generality of multiconfigurational machinery is essential for step-change applications to much larger systems. Linear scaling coupled cluster methods (e.g., using local orbitals) will also surely see greater use in inorganic photochemistry in the future, as will the application of valence bond/coupled cluster hybrid approaches to treat static and dynamic correlations in a compact manner. Such advances are required given the fundamental scientific and technological problems that have at their core a system containing transition metal atom (s) interacting with light.

Acknowledgements We thank the European Research Council for funding under the European Union's Seventh Framework Programme (FP7/2007-2013)/ERC Grant No. 258990.

References

1. Hohenberg P, Kohn W (1964) Inhomogeneous electron gas. Phys Rev 136:B846–B871
2. Kohn W, Sham LJ (1965) Self-consistent equations including exchange and correlation effects. Phys Rev 140:A1133–A1138
3. Pople JA, Gill PMW, Johnson BG (1992) Kohn-Sham density-functional theory within a finite basis set. Chem Phys Lett 199:557–560
4. Moller C, Plesset MS (1934) Note on approximate treatment for many-electron systems. Phys Rev 46:618–622
5. Pople JA, Seeger R, Krishnan R (1977) Variational configuration interaction methods and comparison with perturbation-theory. Int J Quan Chem 149–163
6. Coester F (1958) Bound states of a many-particle system. Nuc phys 7:421–424
7. Coester F, Hummel H (1960) Short-range correlations in nuclear wavefunctions. Nuc phys 17:477–485
8. Kummel H, Luhrmann KH (1972) Equations for linked clusters and energy variational principle. Nuc Phys A A191:525
9. Christiansen O, Koch H, Jorgensen P (1995) Response functions in the CC3 iterative triple excitation model. J Chem Phys 103:7429–7441
10. Christiansen O, Koch H, Jorgensen P (1995) The 2nd-order approximate coupled-cluster singles and doubles model CC2. Chem Phys Lett 243:409–418
11. Taylor DJ, Paterson MJ (2010) Calculations of the low-lying excited states of the TiO_2 molecule. J Chem Phys 133
12. Knowles PJ (1991) A full-CI study of the energetics of the reaction $F + H_2 \rightarrow HF + H$. Chem Phys Lett 185:555–561

13. Roos BO, Taylor PR, Siegbahn PEM (1980) A complete active space SCF method (CASSCF) using a density-matrix formulated super-CI approach. Chem Phys 48:157–173
14. Gagliardi L, Roos BO (2007) Multiconfigurational quantum chemical methods for molecular systems containing actinides. Chem Soc Rev 36:893–903
15. Frisch MJ, Trucks GW, Schlegel HB, Scuseria GE, Robb MA, Cheeseman JR, Montgomery JJA, Vreven T, Kudin KN, Burant JC, Millam JM, Iyengar SS, Tomasi J, Barone V, Mennucci B, Cossi M, Scalmani G, Rega N, Petersson GA, Nakatsuji H, Hada M, Ehara M, Toyota K, Fukuda R, Hasegawa J, shida M, Nakajima T, Honda Y, Kitao O, Nakai H, Klene M, Li X, Knox JE, Hratchian HP, Cross JB, Bakken V, Adamo C, Jaramillo J, Gomperts R, Stratmann RE, Yazyev O, Austin AJ, Cammi R, Pomelli C, Ochterski JW, Ayala PY, Morokuma K, Voth GA, Salvador P, Dannenberg JJ, Zakrzewski VG, Dapprich S, Daniels AD, Strain MC, Farkas O, Malick DK, Rabuck AD, Raghavachari K, Foresman JB, Ortiz JV, Cui Q, Baboul AG, Clifford S, Cioslowski J, Stefanov BB, Liu G, Liashenko A, Piskorz P, Komaromi I, Martin RL, Fox DJ, Keith T, Al-Laham MA, Peng CY, Nanayakkara A, Challacombe M, Gill PMW, Johnson B, Chen W, Wong MW, Gonzalez C, Pople JA (2009) Gaussian 09, Revision A.1, Gaussian, Inc., Wallingford
16. Aquilante F, De Vico L, Ferre N, Ghigo G, Malmqvist P-A, Neogrady P, Pedersen TB, Pitonak M, Reiher M, Roos BO, Serrano-Andres L, Urban M, Veryazov V, Lindh R (2010) Software news and update molcas 7: the next generation. J Comput Chem 31:224–247
17. Shepard R, Shavitt I, Pitzer RM, Comeau DC, Pepper M, Lischka H, Szalay PG, Ahlrichs R, Brown FB, Zhao JG (1988) A progress report on the status of the columbus MRCI program system. Int J Quan Chem 149–165
18. Lischka H, Shepard R, Shavitt I, Pitzer RM, Dallos M, Muller T, Szalay PG, Brown FB, Ahlrichs R, Bohm HJ, Chang A, Comeau DC, Gdanitz R, Dachsel H, Erhardt CME, Hochtl P, Irle S, Kedziora GS, Kovar T, Parasuk V, Pepper MJM, Scharf P, Schiffer H, Schindler M, Schuler M, Seth M, Stahlberg EA, Zhao J-G, Yabushita S, Zhang Z, Barbatti M, Matsika S, Schuurmann M, Yarkony DR, Brozell SR, Beck EV, Blaudeau M, Ruckenbauer M, Sellner B, Plasser F, Szymczak JJ (2012) Columbus, an ab initio electronic structure program
19. Lischka H, Shepard R, Pitzer RM, Shavitt I, Dallos M, Muller T, Szalay PG, Seth M, Kedziora GS, Yabushita S, Zhang ZY (2001) High-level multireference methods in the quantum-chemistry program system columbus: Analytic MR-CISD and MR-AQCC gradients and MR-AQCC-LRT for excited states, guga spin-orbit CI and parallel CI density. Phys Chem Chem Phys 3:664–673
20. Lischka H, Shepard R, Brown FB, Shavitt I, (1981) New implementation of the graphical unitary-group approach for multi-reference direct configuration-interaction calculations. Int J Quan Chem 91–100
21. Lischka H, Mueller T, Szalay PG, Shavitt I, Pitzer RM, Shepard R (2011) Columbus-a program system for advanced multireference theory calculations. Wiley Interdisciplinary Rev Computat Mol Sci 1:191–199
22. Guest MF, Bush IJ, Van Dam HJJ, Sherwood P, Thomas JMH, Van Lenthe JH, Havenith RWA, Kendrick J (2005) The Gamess-UK electronic structure package: algorithms, developments and applications. Mol Phys 103:719–747
23. Andersson K, Malmqvist PA, Roos BO, Sadlej AJ, Wolinski K (1990) 2nd-Order perturbation-theory with a CASSCF reference function. J Phys Chem 94:5483–5488
24. Andersson K, Malmqvist PA, Roos BO (1992) 2nd-Order perturbation-theory with a complete active space self-consistent field reference function. J Chem Phys 96:1218–1226
25. Worth GA, Cederbaum LS (2004) Beyond Born-Oppenheimer: Molecular dynamics through a conical intersection. Ann Rev Phys Chem 55:127–158
26. Althorpe SC, Worth GA (2004) Collaborative computational project on molecular quantum dynamics (CCP6). Daresbury Laboratory, Daresbury
27. van der Kamp MW, Mulholland AJ (2013) Combined quantum mechanics/molecular mechanics (QM/MM) methods in computational enzymology. Biochemistry 52:2708–2728

28. Senn HM, Thiel W (2009) QM/MM methods for biomoleculr systems. Angew Chem Int Ed 48:1198–1229
29. Garino C, Salassa L (2013) The photochemistry of transition metal complexes using density functional theory. Philos Trans R Soc A 371:20120134
30. Daniel C (2003) Electronic spectroscopy and photoreactivity in transition metal complexes. Coord Chem Rev 238:143–166
31. Rosa A, Ricciardi G, Baerends EJ, Stufkens DJ (1995) Density-functional study of ground and excited-states of $Mn_2(CO)_{10}$. Inorg Chem 34:3425–3432
32. van Gisbergen SJA, Groeneveld JA, Rosa A, Snijders JG, Baerends EJ (1999) Excitation energies for transition metal compounds from time-dependent density functional theory. Applications to MnO_4^-, $Ni(CO)_4$, and $Mn_2(CO)_{10}$. J Phys Chem A 103:6835–6844
33. Rosa A, Baerends EJ, van Gisbergen SJA, van Lenthe E, Groeneveld JA, Snijders JG (1999) Electronic spectra of $M(CO)_6$ (M=Cr, Mo, W) revisited by a relativistic TD-DFT approach. J Am Chem Soc 121:10356–10365
34. Pollak C, Rosa A, Baerends EJ (1997) Cr-CO photodissociation in $Cr(CO)_6$: reassessment of the role of ligand-field excited states in the photochemical dissociation of metal-ligand bonds. J Am Chem Soc 119:7324–7329
35. Gray HB, Beach NA (1963) The electronic structure of octahedral metal complexes. I. Metal hexacarbonyls and hexacynanides. J Am Chem Soc 85:2922–2927
36. McKinlay RG, Zurek JM, Paterson MJ (2010) Vibronic coupling in inorganic systems: photochemistry, conical intersections, and the Jahn–Teller and pseudo-Jahn–Teller effects. Adv Inorg Chem 62:351–390
37. McKinlay RG, Paterson MJ (2010) In: Köppel H, Barentzen H, Yarkony DR (eds) Jahn-Teller effect: fundamentals and implications for physics and chemistry. Springer-Verlag, Heidelberg, p 311
38. Aarnts MP, Stufkens DJ, Sola M, Baerends EJ (1997) Coordinative behavior of the CNCN ligand. Experimental and density functional study of spectroscopic properties and bonding in the $Cr(CO)_5CNCN$ complex. Organometallics 16:2254–2262
39. Wilms MP, Baerends EJ, Rosa A, Stufkens DJ (1997) Density functional study of the primary photoprocesses of manganese pentacarbonyl chloride $(MnCl(CO)_5)$. Inorg Chem 36:1541–1551
40. McKinlay RG, Paterson MJ (2012) A time-dependent density functional theory study of the structure and electronic spectroscopy of the group 7 mixed-metal carbonyls: $MnTc(CO)_{10}$, $MnRe(CO)_{10}$, and $TcRe(CO)_{10}$. J Phys Chem A 116:9295–9304
41. Goumans TPM, Ehlers AW, van Hemert MC, Rosa A, Baerends EJ, Lammertsma K (2003) Photodissociation of the phosphine-substituted transition metal carbonyl complexes $Cr(CO)_5L$ and $Fe(CO)_4L$: a theoretical study. J Am Chem Soc 125:3558–3567
42. Neugebauer J, Baerends EJ, Nooijen M (2005) Vibronic structure of the permanganate absorption spectrum from time-dependent density functional calculations. J Phys Chem A 109:1168–1179
43. Gunaratne TC, Gusev AV, Peng XZ, Rosa A, Ricciardi G, Baerends EJ, Rizzoli C, Kenney ME, Rodgers MAJ (2005) Photophysics of octabutoxy phthalocyaninato-Ni(II) in toluene: Ultrafast experiments and DFT/TD-DFT studies. J Phys Chem A 109:2078–2089
44. Rosa A, Ricciardi G, Gritsenko O, Baerends EJ (2004) Excitation energies of metal complexes with time-dependent density functional theory. Struc Bond 112:49–116
45. Kuhn O, Hachey MRD, Rohmer MM, Daniel C (2000) A CASSCF/CASPT2 study of the low-lying excited states of $Mn_2(CO)_{10}$. Chem Phys Lett 322:199–206
46. Zalis S, Ben Amor N, Daniel C (2004) Influence of the halogen ligand on the near-UV-visible spectrum of $Ru(X)(Me)(CO)_2$((alpha-diimine) (X=Cl, L; alpha-diimine=Me-DAB, iPr-DAB; DAB=1,4-diaza-1,3-butadiene)): an ab initio and TD-DFT analysis. Inorg Chem 43:7978–7985
47. Zakrzewski J, Delaire JA, Daniel C, Cote-Bruand I (2004) $W(CO)_5$-pyridine Pi-acceptor complexes: theoretical calculations and a laser photolysis study. New J Chem 28:1514–1519

48. Villaume S, Daniel C (2005) Emission spectroscopy of metal-to-ligand-charge-transfer states of HRe(CO)$_3$(H-DAB), model system for alpha-diimine rhenium tricarbonyl complexes. Comptes Rendus Chimie 8:1453–1460

49. Vallet V, Strich A, Daniel C (2005) Spin-orbit effects on the electronic spectroscopy of transition metal dihydrides H$_2$M(CO)$_4$ (M=Fe, Os). Chem Phys 311:13–18

50. Ambrosek D, Villaume S, Gonzalez L, Daniel C (2006) A multi state-CASPT2 vs. TD-DFT study of the electronic excited states of RCo(CO)$_4$ (R=H, CH$_3$) organometallic complexes. Chem Phys Lett 417:545–549

51. Bossert J, Daniel C (2006) Trans-cis photoisomerization of the styrylpyridine ligand in Re(CO)$_3$(2,2 '-bipyridine)(t-4-styrylpyridine)$^+$: Role of the metal-to-ligand charge-transfer excited states. Chem Eur J 12:4835–4843

52. Ambrosek D, Villaume S, Daniel C, Gonzalez L (2007) Photoactivity and UV absorption spectroscopy of RCo(CO)$_4$ (R=H, CH$_3$) organometallic complexes. J Phys Chem A 111:4737–4742

53. Villaume S, Strich A, Daniel C, Perera SA, Bartlett RJ (2007) A coupled cluster study of the electronic spectroscopy and photochemistry of Cr(CO)$_6$. Phys Chem Chem Phys 9:6115–6122

54. Ben Arnor N, Villaume S, Maynau D, Daniel C (2006) The electronic spectroscopy of transition metal carbonyls: the tough case of Cr(CO)$_6$. Chem Phys Lett 421:378–382

55. Koch H, Jørgensen P (1990) Coupled cluster response functions. J Chem Phys 93:3333–3344

56. Stanton JF, Bartlett RJ (1993) Equation of motion coupled-cluster method: a systematic biorthogonal approach to molecular excitation energies, transition probabilties and excited state properties. J Chem Phys 98:7029–7039

57. Trushin SA, Kosma K, Fuss W, Schmid WE (2008) Wavelength-independent ultrafast dynamics and coherent oscillation of a metal-carbon stretch vibration in photodissociation of Cr(CO)$_6$ in the region of 270–345 nm. Chem Phys 347:309–323

58. Trushin SA, Fuss W, Schmid WE, Kompa KL (1998) Femtosecond dynamics and vibrational coherence in gas-phase ultraviolet photodecomposition of Cr(CO)$_6$. J Phys Chem A 102:4129–4137

59. Trushin SA, Fuss W, Schmid WE (2000) Conical intersections, pseudorotation and coherent oscillations in ultrafast photodissociation of group-6 metal hexacarbonyls. Chem Phys 259:313–330

60. Trushin SA, Fuss W, Kompa KL, Schmid WE (2000) Femtosecond dynamics of Fe(CO)$_5$ photodissociation at 267 nm studied by transient ionization. J Phys Chem A 104:1997–2006

61. Fuss W, Trushin SA, Schmid WE (2001) Ultrafast photochemistry of metal carbonyls. Res Chem Intermed 27:447–457

62. Villaume S, Daniel C, Strich A, Perera SA, Bartlett RJ (2005) Quantum chemical study of the electronic structure of NiCH$_2^+$ in its ground state and low-lying electronic excited states. J Chem Phys 122(4):044313–044316

63. Kayanuma M, Gindensperger E, Daniel C (2012) Inorganic photoisomerization: the case study of rhenium(I) complexes. Dalton Trans 41:13191–13203

64. Kayanuma M, Daniel C, Koppel H, Gindensperger E (2011) Photophysics of isomerizable Re(I complexes: a theoretical analysis. Coord Chem Rev 255:2693–2703

65. Gindensperger E, Koppel H, Daniel C (2010) Mechanism of visible-light photoisomerization of a Rhenium(I) carbonyl-diimine complex. Chem Comm 46:8225–8227

66. Bakova R, Chergui M, Daniel C, Vlcek A Jr, Zalis S (2011) Relativistic effects in spectroscopy and photophysics of heavy-metal complexes illustrated by spin-orbit calculations of Re(imidazole)(CO)$_3$(phen)$^+$. Coord Chem Rev 255:975–989

67. Pellegrin Y, Sandroni M, Blart E, Planchat A, Evain M, Bera NC, Kayanuma M, Sliwa M, Rebarz M, Poizat O, Daniel C, Odobel F (2011) New heteroleptic bis-phenanthroline copper (I) complexes with dipyridophenazine or imidazole fused phenanthroline ligands: spectral, electrochemical, and quantum chemical studies. Inorg Chem 50:11309–11322

68. Seth M, Ziegler T (2005) Calculation of excitation energies of open-shell molecules with spatially degenerate ground states. I. Transformed reference via an intermediate configuration

kohn-sham density-functional theory and applications to d^1 and d^2 systems with octahedral and tetrahedral symmetries. J Chem Phys 123

69. Wang F, Ziegler T (2005) Theoretical study of the electronic spectra of square-planar platinum (II) complexes based on the two-component relativistic time-dependent density-functional theory. J Chem Phys 123

70. Rudolph M, Ziegler T, Autschbach J (2011) Time-dependent density functional theory applied to ligand-field excitations and their circular dichroism in some transition metal complexes. Chem Phys 391:92–100

71. Pierloot K (2005) In: Olivucci M (ed) Computational photochemistry. Elsevier, Heidelberg, pp 279–315

72. Pierloot K, Tsokos E, Vanquickenborne LG (1996) Optical spectra of $Ni(CO)_4$ and $Cr(CO)_6$ revisited. J Phys Chem 100:16545–16550

73. Zurek JM, Paterson MJ (2009) Theoretical study of the pseudo-Jahn–Teller effect in the edge-sharing bioctahedral complex $Mo_{-2}(DXyIF)_2(O_2CCH_3)_2(\mu_2\text{-}O)_2$. Inorg Chem 48:10652–10657

74. Cotton FA, Daniels LM, Murillo CA, Slaton JG (2002) A pseudo-Jahn–Teller distortion in an $Mo_2(\mu_2\text{-}O)_2$ ring having the shortest Mo-IV-Mo-IV double bond. J Am Chem Soc 124:2878–2879

75. Bearpark MJ, Blancafort L, Robb MA (2002) The pseudo-Jahn–Teller effect: a CASSCF diagnostic. Mol Phys 100:1735–1739

76. Zurek JM, Paterson MJ (2010) Photoisomerization in a platinum-amido pincer complex: an excited-state reaction pathway controlled by localized ligand photochemistry. J Phys Chem Lett 1:1301–1306

77. Harkins SB, Peters JC (2006) Unexpected photoisomerization of a pincer-type amido ligand leads to facial coordination at Pt(IV). Inorg Chem 45:4316–4318

78. Zurek JM, Paterson MJ (2012) Photostereochemistry and photoaquation reactions of $Cr(tn)_3^{3+}$: theoretical studies show the importance of reduced coordination conical intersection geometries. J Phys Chem A 116:5375–5382

79. Zurek JM, Paterson MJ (2012) Photoracemization and excited state relaxation through non-adiabatic pathways in chromium (III) oxalate ions. J Chem Phys 137

80. Levine BG, Chaehyuk K, Queenville J, Martinez TJ (2006) Conical intersections and double excitations in time-dependent density functional theory. Mol Phys 104:1039–1051

Struct Bond (2016) 167: 139–162
DOI: 10.1007/430_2014_147
© Springer-Verlag Berlin Heidelberg 2014
Published online: 10 May 2014

d^{10}-ML$_2$ Complexes: Structure, Bonding, and Catalytic Activity

Lando P. Wolters and F. Matthias Bickelhaupt

Abstract Our goal in this chapter is to show how one can obtain a better understanding of the decisive factors for the selectivity and efficiency of catalytically active metal complexes. This ongoing research project has been designated the 'Fragment-oriented Design of Catalysts' and aims at providing design principles for a more rational development of catalysts. To this end, we have performed a series of studies in which we systematically investigate the effect of a specific variation on the reactivity of the catalyst. Thus, we will summarize previous results on not only how the reaction barrier varies when different bonds are activated by palladium, different ligands are attached to palladium but also how different metal centers perform compared to palladium. In a final section, we present a case study on newly obtained results about the effect of adding substituents with different electronegativity to the phosphine ligands at the metal center. A red thread throughout the chapter, and our methodology in general, is the application of the activation strain model of chemical reactivity. This is a predictive model that provides a quantitative relationship between trends in barrier heights and variation of geometric and electronic properties of the reactants.

Keywords Activation strain model · Bond activation · Bond theory · Catalysis · Density functional calculations · Halogenated phosphine ligands · Ligands · Transition metal complexes

L.P. Wolters
Department of Theoretical Chemistry and Amsterdam Center for Multiscale Modeling,
VU University, De Boelelaan 1083, 1081 HV Amsterdam, The Netherlands

F.M. Bickelhaupt (✉)
Department of Theoretical Chemistry and Amsterdam Center for Multiscale Modeling,
VU University, De Boelelaan 1083, 1081 HV Amsterdam, The Netherlands

Institute for Molecules and Materials, Radboud University Nijmegen, Heyendaalseweg 135,
6525 AJ Nijmegen, The Netherlands
e-mail: F.M.Bickelhaupt@vu.nl

Contents

1 Introduction

In an era of growing concern about the human energy demands and its consequences for the environment, the importance of catalysis is evident. One of the
most ubiquitous families of catalytic cycles is that of cross-coupling reactions
[1, 2]. These catalytic cycles can be applied to form carbon–carbon bonds, which
are of paramount importance for the synthesis of pharmaceuticals, as well as
materials. The active catalytic species in these cross couplings is a transition
metal complex, often based on palladium (Scheme 1). The ongoing importance of
palladium-catalyzed cross-coupling reactions has been emphasized by the fact that
it was the topic of the 2010 Nobel Prize in chemistry [3–5].

Cross-coupling reactions are initiated by the oxidative addition of a substrate
(typically an aryl halide) to the ligated palladium complex PdL_n. This first step is
generally considered to be important for both the efficiency and selectivity of the
process, and is therefore widely studied both experimentally [6–10] and theoretically [11–14]. Although this vast body of work has certainly contributed to the
understanding of catalytic reactivity, it is still hard to predict the reactivity of a
complex, due to the many available variables, such as metal center, ligands,
substrate, solvent, and reaction conditions. Not only insight into the effect of all
these variables is required, but also the interplay between all these effects has to be
understood in order to make reliable predictions. Therefore, despite the huge
amount of available literature, the actual catalyst selection process is still often
done through a process based on experience and trial and error. In order to facilitate
this selection process, we employ theoretical chemistry, which allows variation of
one parameter at a time, under strictly controlled conditions and without any
experimental limitation, while simultaneously the added benefit of available analysis tools allows us to explain the observed effects and eventually their interplay.
This strategy of gradually building up insight into the catalytic activity has been
termed the 'Fragment-oriented Design of Catalysts' [15], and aims at allowing

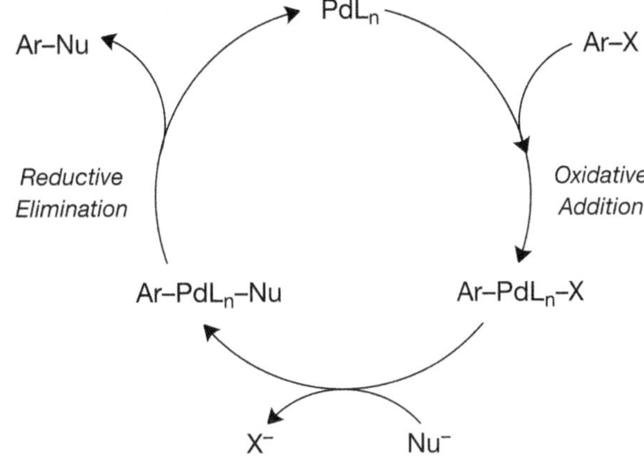

Scheme 1 Schematic catalytic cycle for a palladium-catalyzed cross coupling

chemists in the future to rationally design catalysts with the desired selectivity and optimized efficiency.

In this work, we will briefly discuss a selection of previously obtained insights, either by us or by other groups, as well as newly obtained results on palladium complexes with halogenated phosphine ligands. This case study serves as an instructive example of a study on reactivity, as well as its connection with molecular geometry and bond analyses. To explain the insights, we apply the activation strain model of chemical reactivity [16–19], which we will therefore discuss first. This model, combined with an interaction energy decomposition and qualitative molecular orbital (MO) theory, explains trends in reaction barriers and reaction energies, and it elucidates the bonding mechanism between molecular fragments. Note that this additional insight does not replace but comes on top of a computational quest for the most feasible among several plausible reaction pathways, including unwanted side reactions. Note also that to optimize a catalyst for a specific process, one must consider the complete catalytic cycle, including not only the overall reaction barrier but also the consequences of the stabilization of intermediates for which an energy span model, based on the steady-state approximation, has been developed [20–23].

2   The Activation Strain Model of Chemical Reactivity

The activation strain model of chemical reactivity [16–19] is a fragment-based approach to understand (trends in) chemical reactivity, in terms of the intrinsic properties of reference fragments. Due to its fragment-based nature, the model is most often applied to bimolecular processes, such as oxidative additions [24] as in this work, but also S$_N$2-reactions [16, 25, 26], pericyclic reactions [27], and even barrier-free bond formations, such as hydrogen or halogen bonds [28]. However, also unimolecular processes are successfully studied using the activation strain model [29–31]. Since we are in this work primarily interested in the oxidative-addition reaction, we will discuss the activation strain model as it is most commonly applied to a bimolecular process, with the fragments chosen to be the two reactants: the catalyst and the substrate.

Starting from two isolated reference fragments, the oxidative addition involves the deformation of, and interaction between, the catalyst and the substrate. Within the activation strain model, the relative energy ΔE at a certain point of the reaction energy profile (plotted along the reaction coordinate ζ) is split accordingly into two terms. The first term is the strain energy, $\Delta E_{\mathrm{strain}}$, that is associated with the deformation energy of the fragments from their reference geometries to the geometry they acquire at the point of interest, and eventually also accounting for electronic excitations or relaxations. The second term, the interaction energy ΔE_{int}, accounts for all mutual chemical interactions between these deformed fragments [Eq. (1)].

$$\Delta E(\zeta) = \Delta E_{\mathrm{strain}}(\zeta) + \Delta E_{\mathrm{int}}(\zeta). \tag{1}$$

To obtain insightful results, it is important to choose a proper reaction coordinate. For oxidative additions, the elongation of the bond that is broken has been shown to be a suitable parameter to project the reaction coordinate onto [32]. The activation strain analyses in the present work are therefore projected onto the stretch of the activated bond. The strain energy can be divided into contributions $\Delta E_{\mathrm{strain}}[\mathrm{cat}]$ from the catalyst and $\Delta E_{\mathrm{strain}}[\mathrm{sub}]$ from the substrate. The interaction energy ΔE_{int} can be analyzed in the conceptual framework provided by the Kohn–Sham molecular orbital method [33, 34]. It can be further divided into three physically meaningful terms [Eq. (2)], using a quantitative energy decomposition scheme developed by Ziegler and Rauk [33, 35].

$$\Delta E_{\mathrm{int}}(\zeta) = \Delta V_{\mathrm{elstat}}(\zeta) + \Delta E_{\mathrm{Pauli}}(\zeta) + \Delta E_{\mathrm{oi}}(\zeta). \tag{2}$$

The term $\Delta V_{\mathrm{elstat}}$ corresponds to the classical electrostatic interaction between the unperturbed charge distributions $\rho_A(r) + \rho_B(r)$ of the prepared or deformed fragments A and B that adopt their positions in the overall molecule AB, and is usually attractive. The Pauli repulsion term $\Delta E_{\mathrm{Pauli}}$ comprises the destabilizing interactions between occupied orbitals and is responsible for the steric repulsion (see Fig. 1). This repulsion is caused by the fact that two electrons with the same spin cannot occupy the same region in space. It arises as the energy change associated with the transition from the superposition of the unperturbed electron densities $\rho_A(r) + \rho_B(r)$ of the geometrically deformed but isolated fragments A and B, to the wavefunction $\Psi^0 = N\hat{A}[\Psi_A\Psi_B]$, that properly obeys the Pauli principle through explicit antisymmetrization ($\hat{A}$ operator) and renormalization (N constant) of the product of fragment wavefunctions (see [33] for an exhaustive discussion). The orbital interaction ΔE_{oi} accounts for electron pair bond formation, charge transfer (interaction between occupied orbitals on one fragment with unoccupied orbitals on the other fragment, including the HOMO–LUMO interactions), and

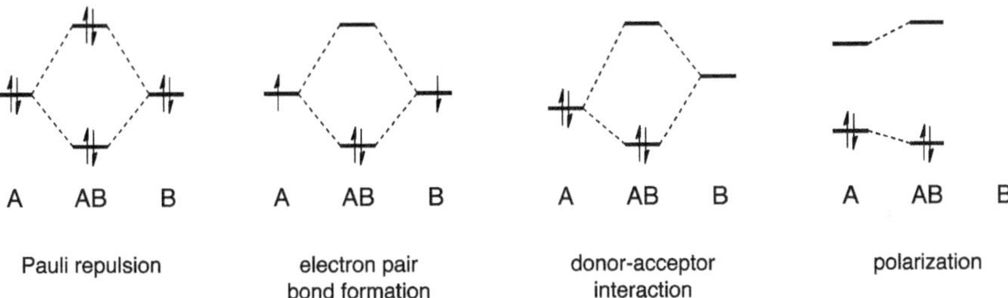

Fig. 1 Schematic representations of Pauli repulsion, and some commonly encountered interactions (electron pair bond formation, donor–acceptor interactions and polarization) contributing to the orbital interaction energy

polarization (empty–occupied orbital mixing on one fragment due to the presence of another fragment). Schematic representations of these interactions are shown in Fig. 1. The orbital interaction term can be further divided into contributions from each irreducible representation Γ of the interacting system [see Eq. (3)].

$$\Delta E_{\text{oi}}(\zeta) = \sum_{\Gamma} \Delta E_{\text{oi}}^{\Gamma}(\zeta). \tag{3}$$

When applied to dispersion-corrected computations, Eq. (2) can be augmented with a term ΔE_{disp}.

With regard to the activation strain model of chemical reactivity and the interaction energy decomposition analyses, it should be noted that the former can be applied using any quantum chemistry software package, whereas the quantitative energy decomposition scheme is a unique feature of the Amsterdam Density Functional program package (ADF) [36–38]. The PyFrag program has been developed as a wraparound for ADF, to streamline performing the activation strain analyses [39].

3 Fragment-Oriented Design of Catalysts

3.1 Activation of Different Bonds

In order to understand the activity of a catalyst, one must first have a decent understanding of the reaction mechanism. To this end, studies have been performed on the activation of a number of different bonds by the most simple model catalyst: a bare Pd(0) atom. This enables one to understand the intrinsic reactivity of the metal atom and how the reaction barrier for oxidative addition is influenced by the different bonds to be activated [17, 18, 40–43]. It was found that the $d^{10}s^0$ occupation of the metal is of great importance for the oxidative-addition reaction, because charge donation from the occupied metal d orbitals into the substrate's σ* orbital is

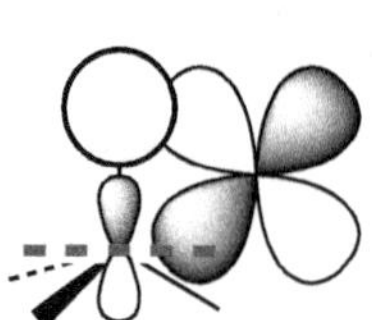 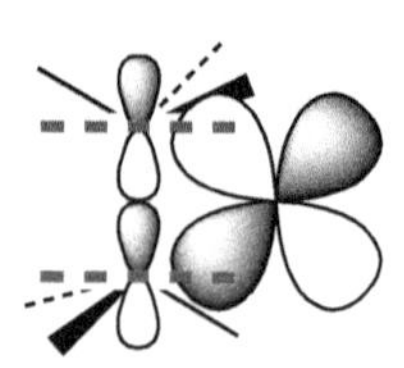 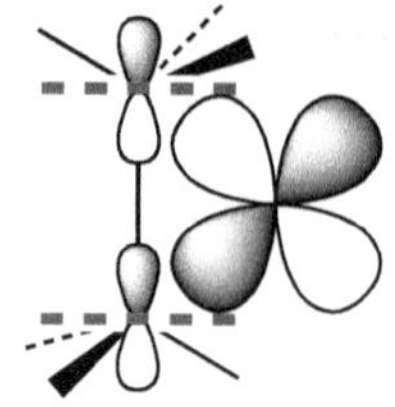

Fig. 2 Different overlap situations for the metal d orbital with a C–H bond (*left*), a C–C bond (*middle*), and with a stretched C–C bond (*right*)

the driving force behind the reaction, while the empty metal s orbital allows the substrate to coordinate to the metal center. To explain the reactivity of Pd(0) towards different bonds, the bond dissociation energy of these bonds is, of course, one of the key factors determining the height of the reaction barrier. For example, addition of halomethanes CH_3X with X=F, Cl, Br, I and At goes with decreasing reaction barriers from F to At [42]. This trend results directly from the decreasing C–X bond strength that is observed in CH_3X along this series.

However, the strength of the bond to be activated is not the only factor that determines the height of the reaction barrier. It is known, for example, that activation of the stronger methane C–H bond can be kinetically more feasible than activation of a weaker ethane C–C bond [44–47]. This has been attributed to the different compositions of the antibonding σ* acceptor orbital [43, 45, 48]. For the C–H bond, this orbital is the antibonding combination of the methyl sp^3 lobe and the 1s orbital on hydrogen, whereas for the C–C bond it is the antibonding combination of two methyl sp^3 lobes. The latter combination has an additional nodal plane, and the C–C bond therefore has to stretch further in order to achieve favorable overlap (avoid cancellation of overlap) with the metal d orbitals (Fig. 2). This leads to a delay in the buildup of stabilizing catalyst–substrate interactions and therefore a destabilization of the transition state. In the activation strain model, this shows up as a ΔE_{int} term that starts to become more stabilizing only at a later stage, while the destabilizing term ΔE_{strain} is already increasing considerably at an early stage of the reaction. A similar effect is found for carbon–halogen bonds C–X, where the σ* orbital is composed of an antibonding combination of the methyl sp^3 hybrid and the halogen p orbital. Furthermore, for ethane C–C activation, the strain energy is additionally destabilized because of the need to bend two methyl groups away in order to make room for the metal, instead of only one methyl group in the case of methane.

3.2 The Effect of Ligand Variation

A logical next step towards more applicable reactions is to study the effect of adding ligands to the palladium center. Firstly, we will discuss the effect of adding a

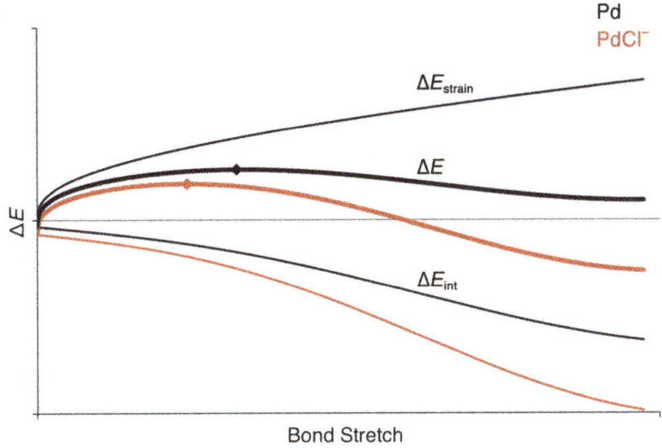

Fig. 3 Schematic comparison of the activation strain analyses obtained for bond activation by Pd(0) (*black*) and anion-assisted PdCl⁻ (*red*). For clarity, the strain curves are coincident, which is a valid approximation for most practical applications

chloride ligand [17]. The effect of adding an anion is known to be able to speed up oxidative-addition reactions to zerovalent palladium complexes, as well as, for example, the rate-determining step in the industrially important Monsanto process [49–53]. By directly comparing activation of H–H, C–H, C–C, and C–Cl bonds by Pd as well as PdCl⁻, such anion assistance was shown to decrease all oxidative addition barriers. Expectedly, adding a chloride anion to the palladium center has not much effect on the strain energy (ΔE_{strain}) curve, which for all reactions closely resembles the bond dissociation profile. The reason for the lower reaction barriers is a substantially stabilized catalyst–substrate interaction ΔE_{int} (Fig. 3). This strengthening becomes larger as the reaction proceeds, because the inherent strength of the catalyst–substrate interaction also increases along the reaction profile. Therefore, the interaction energy profile descends more steeply, shifting the position of the TS to the left. This implies that the transition state is more reactant-like as the exothermicity of the reaction is increased, which is reminiscent of the Hammond postulate [54]. Thus, the origin and mechanism of this postulate emerges naturally in terms of the activation strain model. A further quantitative decomposition of the interaction energy [see Sect. 2 and Eq. (2)] along the energy profile shows that the stronger interaction results from a less destabilizing Pauli energy term, which constitutes the repulsive interactions between occupied orbitals on the fragments (see Fig. 1). These interactions are weakened due to the greater orbital energy gap between the occupied orbitals on PdCl⁻ and those on the substrate. The orbitals of PdCl⁻ are pushed up in energy due to the presence of the negatively charged chloride. The higher occupied catalyst orbitals also strengthen the backbonding interactions to the substrate, but this contribution is masked by the decreased substrate-to-metal donation because also the catalysts' acceptor orbitals are destabilized.

The situation changes considerably when dicoordinated species are studied, such as the bisphosphine complex Pd(PH$_3$)$_2$. Compared to bare Pd or monocoordinated palladium complexes, the barriers for oxidative addition to dicoordinated complexes are significantly higher [55–59]. Activation strain analyses revealed [57] that the main reason for the higher reaction barriers is an increased catalyst strain energy

$\Delta E_{\text{strain}}[\text{cat}]$, that results from the need of bending the ligands away to avoid steric repulsion with the substrate. This destabilizing effect due to the need to decrease the ligand–metal–ligand angle (the bite angle) can be avoided by bending the catalyst in advance, as, for example, in chelate complexes $Pd[PH_2(CH_2)_nPH_2]$ where the bite angle can be controlled by varying the length of the carbon-chain $(CH_2)_n$. It is well known that the reactivity of a catalytic complex depends on its bite angle [21, 60–64]. By studying oxidative addition of several bonds to this series of model catalysts with $n = 2$–6, bite angles from 98° to 156° can be achieved, and a clear relationship was found between the ligand–metal–ligand angles and the activation barriers. Upon decreasing the length of the carbon-chain in the bidentate ligand, the bite angle decreases, and concomitantly the activation barrier is lowered as a result of the less strong destabilizing catalyst strain energy $\Delta E_{\text{strain}}[\text{cat}]$, because there is less need to deform the catalyst. In other words: part of the strain energy is taken out of the reaction energy profile by building it into the catalytically active complex. This steric mechanism is the main reason for the lower reaction barriers for oxidative addition to catalytic complexes with smaller bite angles. It is, however, accompanied by a small electronic effect, namely the enhanced backbonding from a destabilized palladium d orbital to the substrate σ* orbital. In the past it has been suggested by Hofmann and co-workers that this electronic nature was the main reason for the lower barriers for catalysts with smaller bite angles [65]. Although this electronic effect contributes to their increased reactivity, it is marginal compared to the steric effect.

3.3 The Effect of Metal Variation

Yet another parameter to be varied is the metal center itself. Although palladium is widely used, it is not used exclusively. Examples are the rhodium-based catalyst in the already mentioned Monsanto process, but also nickel, platinum, and even gold complexes have been found to be active species [6, 10, 12, 66–68]. In our group, we have studied the activation of C–H, C–C, C–F, and C–Cl bonds by the bare coinage metal cations Cu^+, Ag^+, and Au^+ and compared the results to those of Pd [69]. It was found that in general the second-row elements Pd and Ag^+ have the highest barrier towards bond activation. Even though the barriers for addition to the group 11 cations can be rather similar to the barriers for palladium, the bonding mechanism is different. Upon comparing the activation strain analyses for palladium with those of the group 11 cations (Fig. 4), three observations are easily made: firstly, and expectedly, the strain curves hardly change as there is no $\Delta E_{\text{strain}}[\text{cat}]$ and $\Delta E_{\text{strain}}[\text{sub}]$ is not significantly influenced by the metal center. Secondly, the interaction curves for the group 11 cations start at a lower energy, but are more shallow than for palladium. Thirdly, and as a result of the more shallow interaction energy curves, the reactant side of the energy profile for the cations is stabilized compared to palladium, while the product side is less stabilized or even destabilized, thereby shifting the TS to a more product-like geometry (Fig. 4).

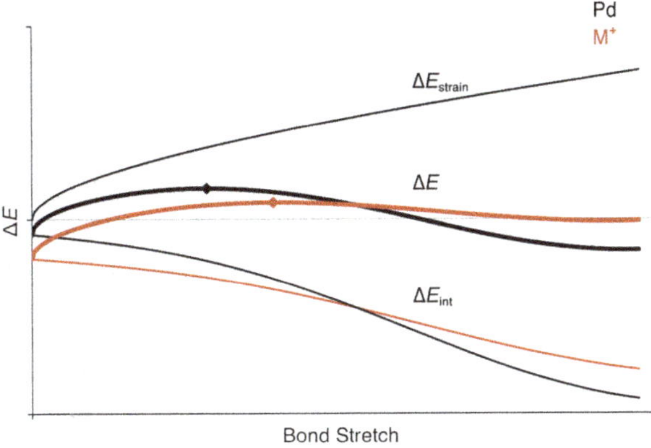

Fig. 4 Schematic comparison of the activation strain analyses obtained for bond activation by Pd(0) (*black*) and a bare cation M$^+$ (*red*). For clarity, the strain curves are coincident, which is a valid approximation for most practical applications

The reason for the more shallow interaction energy curves for the group 11 cations is easy to understand when one considers the different bonding interactions during the oxidative addition process. At the initial stage, the dominant interaction is donation from the substrate into the empty metal s orbital, while at a later stage backdonation from the metal d orbitals into the substrate σ* orbital becomes the most important bonding interaction. The cations, due to their net positive charge, are excellent electron acceptors, which enhances the bonding interactions in the initial stage. For the same reason, they are less good electron donors, resulting in a weaker backbonding interaction at a later stage.

These results are confirmed by a recent follow-up study on the addition of the methane C–H bond to a more extensive set of d^{10} transition metal complexes [70]. In this study, all transition metals surrounding palladium in the periodic table were included, as well as mono- and dicoordinated complexes ML and ML$_2$, where L is either NH$_3$, PH$_3$, or CO. In this work, all reaction profiles were analyzed with respect to the metal in its d^{10}s^0 configuration, which means that the catalysts based on the group 9 metal centers Co, Rh, and Ir are negatively charged, whereas the catalysts based on the group 11 metals are, as in the previous study, positively charged. For some catalysts the d^{10}s^0 configuration (or d^{10}s^0-like when ligands are present) is an excited state, but using the same electronic configuration for all complexes allowed us to make a consistent comparison of the metal centers. Furthermore, this d^{10}s^0 or d^{10}s^0-like configuration corresponds to the ground state of most catalysts used in practice.

Upon comparison of the energy profiles of, for example, Rh(PH$_3$)$_2$$^-$, Pd(PH$_3$)$_2$, and Ag(PH$_3$)$_2$$^+$, it was found that the methane addition barrier generally increases from anionic to cationic complexes. Activation strain analyses, a further interaction energy decomposition and detailed molecular orbital analyses revealed that this is due to poorer backbonding capabilities of the catalytic complex, which lead to a less stabilizing interaction energy term, while there is less variation among the strain energy curves. Furthermore, it was found that also in the group 9 and group 10 triads of metals the catalysts based on the second-row transition metal generally have the highest barrier towards methane activation. Again, the strain energy curves only showed minor differences, and the main reason for this trend is a similar trend in the interaction energy curves. Thus, the catalysts with first- and third-row transition metal centers both have a more stabilizing interaction with the substrate than the second-row transition metal, albeit for different reasons. Compared to the second-row metal centers, the first row metals typically have higher-energy d orbitals (when comparing the relevant $d^{10}s^0$ configurations), which results in stronger backbonding capabilities and therefore lower energy profiles. The second and third row transition metal centers, on the other hand, have similar d orbital energies, but the latter have larger orbitals which show greater overlap with the substrate σ^* orbital, leading to slightly improved backbonding. This is accompanied by the relativistic stabilization of the third row metal $(n+1)$s orbital, which results in better electron-accepting capabilities of the catalytic complex and hence a stronger substrate-to-metal donation.

The differences in catalyst strain energy that were found along this series, as well as the series from $Rh(PH_3)_2^-$ to $Ag(PH_3)_2^+$, resulted from variations in flexibility of the ligand–metal–ligand angle of the catalyst, as has been briefly touched upon before [57]. For example, from $Rh(PH_3)_2^-$ to $Pd(PH_3)_2$ to $Ag(PH_3)_2^+$ the decreased flexibility contributes to the higher barriers. We found that many catalysts have very flat potential energy surfaces for bending the bite angle, and in fact, some ML_2 complexes even have nonlinear equilibrium geometries. This leads us to additional research [71] on the geometries and bonding mechanism of such ML_2 complexes, the results of which are discussed in the next section.

3.4 Nonlinear d^{10}-ML_2 Transition Metal Complexes

In general, d^{10}-ML_2 transition metal complexes are expected to have linear L–M–L angles [72–75], which can be easily rationalized in terms of several models, among which the Walsh diagrams based on MO theory [76]. The main bonding interaction is considered to be σ donation from the ligand lone-pair orbitals into the empty metal $(n+1)$s atomic orbital, which has a bond overlap that is independent of the ligand–metal–ligand angle. The geometries of ML_2 complexes are therefore expected to be linear, due to steric interactions between the ligands, pushing them as far apart as possible. However, nonlinear geometries are observed for some complexes, such as $Ni(CO)_2$ [77, 78]. In a study on a large set of d^{10}-ML_2 complexes, with $M = Co^-$, Rh^-, Ir^-, Ni, Pd, Pt, Cu^+, Ag^+ Au^+ and $L = NH_3$, PH_3 or CO, we have shown that the equilibrium geometries deviate increasingly from linearity when the ligand is a better π-acceptor, and also when the metal is a better electron donor [71]. Thus, from NH_3- to PH_3- to CO-ligated catalysts, the nonlinearity increases, as, for example, shown by $Rh(NH_3)_2^-$ (180°), $Rh(PH_3)_2^-$ (141°), and $Rh(CO)_2^-$ (131°). Furthermore, as along the iso-electronic series of complexes from $Ag(CO)_2^+$ to $Pd(CO)_2$ to $Rh(CO)_2^-$ the electron-donating capability of the metal center increases, the nonlinearity increases as well: whereas $Ag(CO)_2^+$ is linear, $Pd(CO)_2$ has an angle of 156° and $Rh(CO)_2^-$ of 131°.

Fig. 5 Schematic representation of the π-backbonding interactions in d^{10}-ML$_2$ complexes at 180° and 90°. Figure adapted from [71]

A detailed analysis of the bonding mechanism of these intrinsically bent catalytic complexes has shown that π-backbonding plays a critical role in arriving at this geometric preference, as suggested by the two trends just described. Thus, we have analyzed the bonding mechanism between a monocoordinated ML fragment and the second ligand L′ while varying the L–M–L′ angle from 180° to 90°. It was found that, upon bending, the bonding interactions between LM and L′ become stronger due to increased π-backbonding, while simultaneously, as expected, the steric repulsion between the fragments strengthens as well. The reason for this enhanced π-backbonding is easy to understand in terms of the changing orbital interactions between the fragments as the L–M–L′ angle is bent: in the linear geometry, the two degenerate π-accepting orbitals on L′ interact with d-derived orbitals on LM that are already stabilized due to π-backbonding within LM. When the angle is decreased, the overlap between one π* orbital on L′ and the bonding orbital on LM is decreased, while simultaneously this same π* orbital builds up overlap with a non-bonding, essentially pure d orbital on the metal fragment (Fig. 5). This latter orbital has no bonding interactions with the first ligand L and is therefore higher in energy, which favors the π-backdonation to the second ligand L′. If this additional stabilization is stronger than the increased steric repulsion, the minimum on the energy profile shifts towards angles smaller than 180°, leading to nonlinear equilibrium geometries.

This gain in stabilization is strongest when the intrinsic π-backbonding is strongest, that is, when the metal is a better electron donor, or when the ligand is a better π-acceptor. Therefore, the catalytic complexes are more strongly bent along Ag(CO)$_2^+$, Pd(CO)$_2$, and Rh(CO)$_2^-$ and along Rh(NH$_3$)$_2^-$, Rh(PH$_3$)$_2^-$, and Rh(CO)$_2^-$.

Furthermore, we also noted that going down the periodic table, for example from Ni(CO)$_2$ to Pd(CO)$_2$ to Pt(CO)$_2$, the complexes become more linear. This is attributed to, firstly, a weaker π-backbonding for Pd(CO)$_2$ than for Ni(CO)$_2$, because the palladium 4d orbitals are lower in energy than the nickel 3d orbitals (again, analyses are relative to the d^{10}s^0 configuration, which is not the atomic ground state for Ni). Secondly, from Pd(CO)$_2$ to Pt(CO)$_2$ the steric repulsion upon bending becomes stronger due to a stronger admixture of the relativistically stabilized platinum 6s orbital with the occupied d$_{z^2}$ orbital on the ML fragment. This leads to an enlarged torus of the d$_{z^2}$ orbital, and upon bending towards 90°, an increased repulsive overlap of this torus with the lone pair on the ligand.

Although the arguments presented here are developed for d^{10}-ML_2 complexes, similar arguments account for the nonlinear structures observed for d^0 metal complexes with π-donating ligands [79–83]. Furthermore, similar reasoning is expected to account for a related effect described for d^8-$M(CO)_2 L_2$ complexes by Eisenstein and Caulton, where strong π-accepting ligands induce non-planarity [84, 85].

4 Case Study: Halogenated Phosphine Ligands at Palladium

4.1 Introduction

In this section we will present newly obtained results on the activation of the methane C–H bond by halogen-substituted palladium-phosphine complexes $Pd(PX_3)_2$ where X=F, Cl, Br, or I. This topic has been briefly touched upon in previous work by us [57] as well as others [55], but these studies only included $Pd(PCl_3)_2$. Here, we will therefore discuss not only the difference in reactivity upon going from $Pd(PH_3)_2$ to halogen-substituted $Pd(PX_3)_2$ but also the effect of decreasing electronegativity along $Pd(PF_3)_2$, $Pd(PCl_3)_2$, $Pd(PBr_3)_2$, and $Pd(PI_3)_2$, which is new. Furthermore, as the bulkiness of the ligands increases from PH_3 to PF_3, PCl_3, PBr_3, and PI_3, we do not only expect electronic effects to play a role, but steric effects as well. Thus, by studying the reactivity of this series of catalysts, we investigate both electronic and steric effects on catalytic activity. However, because in a previous study [57] the $Pd(PCl_3)_2$ complex was found to have a nonlinear equilibrium geometry (a feature that was overlooked by Fazaeli and co-workers [55]),[1] we will first perform detailed bonding analyses to investigate the reasons behind this nonlinearity of $Pd(PCl_3)_2$ and compare its situation to the other halogenated phosphine-catalysts in this series. Interestingly, the latter appear to have nonlinear geometries as well. Furthermore, we will also compare these findings to the results discussed in Sect. 3.4.

4.2 M–L Bonding Analysis and Pd(PX₃)₂ Geometries

Our dispersion-corrected computations at ZORA-BLYP-D3/TZ2P (the computational details are described in [62], except that we have now also included dispersion corrections using Grimme's third-generation DFT-D3 method, as described in [86]) revealed that all halogen-substituted bisphosphine palladium complexes

---

[1] We have performed a geometry optimization of $Pd(PCl_3)_2$ at the computational level described in reference 49, resulting in a P–Pd–P angle of 135.5°. We find that the linear conformer is 1.4 kcal mol^{-1} higher in energy, with two degenerate imaginary frequencies, both corresponding to bending the complex.

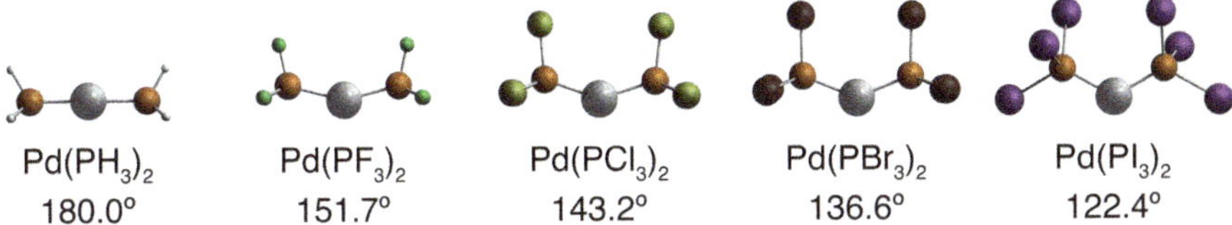

Fig. 6 Equilibrium geometries and P–Pd–P angles of hydrogen- and halogen-substituted palladium-phosphine complexes. Computed at dispersion-corrected ZORA-BLYP-D3/TZ2P

Pd(PX$_3$)$_2$ have nonlinear geometries. Initially, one may expect that the complexes become more linear from Pd(PF$_3$)$_2$ to Pd(PI$_3$)$_2$, based on stronger steric repulsions between the heavier halogens. We find, however, that the opposite is true: along this series the P–Pd–P angle in the equilibrium geometries decreases from 151.7° for X=F to 143.2° (X=Cl), 136.6° (X=Br), and 122.4° for X=I, as shown in Fig. 6. Furthermore, we find that Pd(PH$_3$)$_2$, Pd(PF$_3$)$_2$, Pd(PCl$_3$)$_2$, and Pd(PBr$_3$)$_2$ have eclipsed geometries, leading to a D_{3h}-symmetric geometry for Pd(PH$_3$)$_2$ and C_{2v}-symmetric geometries for the halogenated Pd(PX$_3$)$_2$ complexes. In the latter, two halogens from different ligands point towards each other. For Pd(PI$_3$)$_2$, we find that the ligands are rotated, avoiding close contacts between the iodines on one ligand with the iodines on the other ligand, lowering the symmetry of the complex to C_2.

It is tempting to attribute the bending of these complexes to stronger dispersion interactions between the heavier halogen substituents on different ligands, basically pulling the ligands towards each other. However, dispersion-free computations at the otherwise same ZORA-BLYP/TZ2P level reveal that also without dispersion the angles decrease steadily from 153.9° for Pd(PF$_3$)$_2$ to 141.6° for Pd(PI$_3$)$_2$ (results not shown). Thus, although dispersion contributes to the effect, it is not exclusively responsible for this nonlinearity.

We recall from previous work that sufficient π-backbonding can lead to nonlinear ML$_2$ geometries, because, upon bending from 180° to 90°, this π-backbonding is enhanced (see [71], as summarized in Sect. 3.4). In order to investigate the π-accepting properties of the ligands included in this work, we have performed a bond energy decomposition for the Pd–L bond in monocoordinated PdPH$_3$ and PdPX$_3$ (Table 1). We find a Pd–PH$_3$ bond energy of -40.9 kcal mol^{-1}, whereas the halogen-substituted phosphines bind a little stronger to Pd, with bonding energies between -43.8 and -46.1 kcal mol^{-1}. A further decomposition using Eqs. (1) and (2) reveals that, indeed, the halogen-substituted phosphines have a larger contribution from the π-component of ΔE_{oi} and hence are apparently better π-acceptors. This also follows from the lower π* orbital energies which decrease from -1.5 to -2.4, -2.7, and -3.0 eV from PF$_3$ to PI$_3$, which are all lower than that of PH$_3$ at -0.2 eV. However, we do not find stronger π-backbonding along the halogen-substituted series. The reason is that the π* orbital on PX$_3$ (which has antibonding character between P and X) is increasingly localized on the halogen substituents, and less on the phosphorus atom. Therefore, the overlap between the π* orbital and the Pd d orbitals decreases along this series, thereby counteracting the effect of the lower energy of the π* orbital. Thus, while the stronger π-backbonding for the halogen-substituted phosphines may explain

Table 1 Metal–ligand bond energies and decomposition [Eq. (2)] in kcal mol^{-1} of the monoligated PdPX$_3$ complexes

	ΔE	ΔE_{strain}	ΔE_{int}	ΔE_{disp}	ΔV_{elstat}	ΔE_{Pauli}	ΔE_{oi}	ΔE_{oi}^{σ}	ΔE_{oi}^{π}
Pd–PH$_3$	−40.9	+0.3	−41.2	−1.4	−165.6	+189.4	−63.6	−35.2	−28.4
Pd–PF$_3$	−44.4	+0.1	−44.5	−2.4	−157.8	+193.4	−77.7	−36.5	−41.3
Pd–PCl$_3$	−43.8	+0.1	−43.9	−5.1	−141.8	+178.8	−75.9	−34.4	−41.5
Pd–PBr$_3$	−45.2	+0.4	−45.6	−6.2	−133.7	+172.0	−77.7	−36.1	−41.6
Pd–PI$_3$	−46.1	+0.6	−46.7	−7.0	−131.0	+169.0	−77.7	−36.8	−40.9

Computed at dispersion-corrected ZORA-BLYP-D3/TZ2P

why the Pd(PX$_3$)$_2$ complexes are bent whereas Pd(PH$_3$)$_2$ is not, it does not explain the increased nonlinearity from Pd(PF$_3$)$_2$ to Pd(PI$_3$)$_2$.

We have also performed bonding analyses between ML and the second ligand, whereby we start from D_{3h}-symmetric Pd(PH$_3$)$_2$ or Pd(PX$_3$)$_2$ complexes optimized at the dispersion-free ZORA-BLYP/TZ2P level, and then bend the complex from 180° to 90° without further optimization. This way, we eliminate any geometric relaxation effects, allowing for a concise, but detailed investigation of the bonding mechanism. Thus, in Fig. 7 we show the results of the interaction energy decomposition for Pd(PH$_3$)$_2$ as well as the series of Pd(PX$_3$)$_2$ complexes. As this graph reveals, there is no significant difference between the orbital interaction curves within the halogen-substituted series. Also a further decomposition of this term into contributions from each respective irreducible representation (Eq. (3); results not shown) does not reveal any factor contributing significantly to a preference for nonlinear geometries. Figure 7 does reveal, however, that the minimum on the energy profiles shifts to smaller L–M–L angles from Pd(PH$_3$)$_2$ to Pd(PF$_3$)$_2$, and further to Pd(PI$_3$)$_2$, because the Pauli repulsion increases less steeply for the Pd(PX$_3$)$_2$ series than for Pd(PH$_3$)$_2$, and also along the Pd(PX$_3$)$_2$ series as the halogen substituents become heavier.

This, again counterintuitive, trend originates from the composition of the highest occupied MOs (HOMOs) on the L and ML fragments. The HOMO on the ligand L is the lone pair on phosphorus. For PH$_3$, this is the bonding combination of the hydrogen s orbitals and the phosphorus p$_z$ orbital (with antibonding admixture of the phosphorus s orbital), which is strongly localized on phosphorus (Fig. 8). For the halogenated PX$_3$ ligands, the HOMO has considerably more admixture of the substituent halogen orbitals. It consists of the p$_z$ orbitals on P and X, mixing in antibonding fashion. Thus, the larger amplitude is on the more electropositive phosphorus atom. As from F to I the halogen becomes less electronegative, this orbital becomes less localized on phosphorus (Fig. 8).

For PdPH$_3$ and PdPX$_3$, the HOMO is the antibonding combination of the ligand lone pair and the d$_{z^2}$ orbital on Pd. Because from PH$_3$ to PF$_3$, PCl$_3$, PBr$_3$, and PI$_3$ the ligand lone pair becomes less localized on phosphorus, there is less destabilization from repulsions between this orbital and the palladium d$_{z^2}$ orbital. Due to the decreased destabilization, there is less Pd 5s admixture in the PdPH$_3$ or PdPX$_3$ HOMO, resulting in the torus of the Pd d$_{z^2}$ orbital becoming smaller (Fig. 8). It is this smaller torus from PdPH$_3$ to PdPX$_3$, and from PdPF$_3$ to PdPI$_3$, combined with the lone-pair orbital on the second ligand being less localized on phosphorus, that

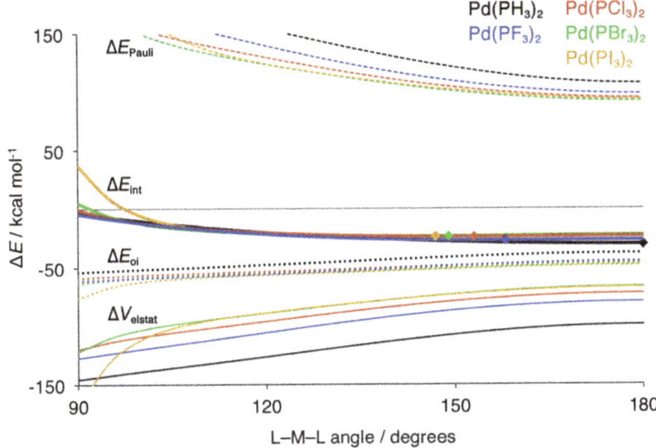

Fig. 7 Bond energy decomposition [Eq. (2)] along the L–M–L angles for Pd(PH$_3$)$_2$ and halogen-substituted Pd(PX$_3$)$_2$ complexes. *Dots* indicate the position of the minimum on the energy profile. Due to the use of frozen geometries and the omission of dispersion corrections, all minima are shifted to the right. Computed at ZORA-BLYP/TZ2P

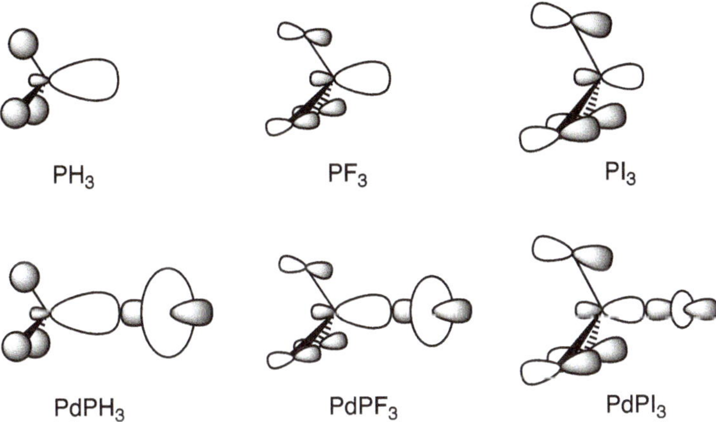

Fig. 8 Schematic representations of the HOMO on PH$_3$, PF$_3$, and PI$_3$ (*top*, from *left* to *right*) and on monocoordinated PdPH$_3$, PdPF$_3$, and PdPI$_3$ (*bottom*, from *left* to *right*)

results in a less steeply increasing Pauli repulsion term for the halogen-substituted catalysts compared to Pd(PH$_3$)$_2$, as well as along the Pd(PX$_3$)$_2$ series as the halogen becomes heavier (Fig. 7)

4.3   Reactivity Towards the Methane C–H Bond

For all catalytic complexes the methane C–H bond activation starts from a reactant complex (RC) that is more strongly bound along the series of catalysts, from -1.9 kcal mol^{-1} for Pd(PH$_3$)$_2$ to -4.7 kcal mol^{-1} for Pd(PI$_3$)$_2$ (Table 2 and

Table 2 Relative energies (kcal mol^{-1}) of the stationary points and transition states for methane C–H activation by the different palladium-based catalysts

	RC	TS	PC
Pd(PH$_3$)$_2$	−1.9	+29.5	+24.3
Pd(PF$_3$)$_2$	−2.8	+26.6	+24.1
Pd(PCl$_3$)$_2$	−3.7	+24.1	+22.1
Pd(PBr$_3$)$_2$	−4.4	+23.7	+22.3
Pd(PI$_3$)$_2$	−4.7	+25.1	+24.1

Computed at dispersion-corrected ZORA-BLYP-D3/TZ2P

Fig. 9). This is because, along this series, the catalytic complexes are bent further and therefore have a sterically less shielded metal center, allowing for a stronger substrate–catalyst interaction immediately at the beginning of the reaction. This interaction is further strengthened by increasingly stabilizing dispersion interactions between methane and the catalyst complexes with the heavier halogens.

As the reaction proceeds, a transition state (TS) is encountered at +29.5 kcal mol^{-1} for Pd(PH$_3$)$_2$, and at slightly lower energies for the Pd(PX$_3$)$_2$ series, in line with findings of previous studies [55, 57]. Along the Pd(PX$_3$)$_2$ series, the barriers first decrease from +26.6 kcal mol^{-1} for Pd(PF$_3$)$_2$ to +24.1 kcal mol^{-1} for Pd(PCl$_3$)$_2$ and +23.7 kcal mol^{-1} for Pd(PBr$_3$)$_2$, and then increase again to +25.1 kcal mol^{-1} for Pd(PI$_3$)$_2$.

Based on activation strain analyses along part of the reaction energy profile obtained by the Transition-Vector Approximation to the IRC (TV-IRC) [32], we find that this ordering of the barriers is the result of two counteracting trends (Fig. 10), namely a reduced strain energy from the hydrogen to the halogen substituents, and a further reduction when the halogens become heavier. The second trend is a simultaneous weakening of the interaction between the catalyst complex and methane substrate (which we address later on). Because the strain energy from Pd(PH$_3$)$_2$ to Pd(PF$_3$)$_2$ and onwards to Pd(PI$_3$)$_2$ decreases in progressively smaller steps, while, on the other hand, the interaction energy terms weaken with progressively larger steps, the oxidative addition barrier first decreases from Pd(PH$_3$)$_2$ to Pd(PBr$_3$)$_2$ and then increases again to Pd(PI$_3$)$_2$.

A further decomposition of the strain energy into individual contributions from the catalyst and substrate clearly reveals that the differences in catalyst strain are decisive. These differences are directly related to the flexibility, or indeed nonlinearity, of the complexes. Thus, although the easier bending of the L–M–L angle contributes to the progressively decreasing catalyst strain from Pd(PH$_3$)$_2$ to Pd(PI$_3$)$_2$, the potential energy surfaces for bending these complexes are very flat. The bending itself therefore only contributes a few kcal mol^{-1} to the total catalyst strain. A significant contribution to the catalyst deformation energy stems from further tilting and rotation of the ligands, which accompanies the bending. These deformations are less needed when the L–M–L angle is intrinsically more bent, and therefore add to the lowering of the catalyst strain originating from the increased flexibility of the catalyst. From Pd(PBr$_3$)$_2$ to Pd(PI$_3$)$_2$, however, this increase in flexibility is less important because it has reached a point where the catalysts are flexible enough, and the direct steric interaction between the ligands prevents further bending. This steric repulsion is also revealed in Fig. 7 by the strong

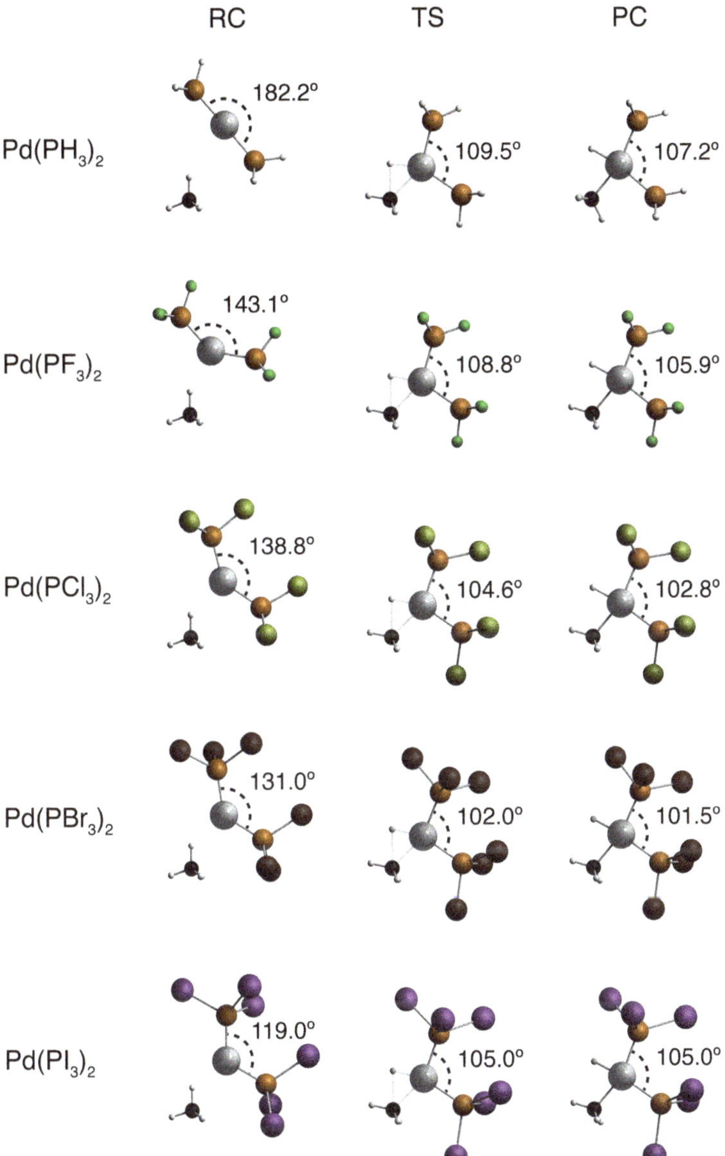

Fig. 9 Geometries of the RC, TS, and PC along the energy profile for oxidative addition of methane to Pd(PH$_3$)$_2$, Pd(PF$_3$)$_2$, Pd(PCl$_3$)$_2$, Pd(PBr$_3$)$_2$, and Pd(PI$_3$)$_2$. Computed at dispersion-corrected ZORA-BLYP-D3/TZ2P

increase in Pauli repulsion that occurs at angles below 110°. Although only the beginning of this sharp increase of the Pauli repulsion term is visible in the graph and most of it is off the scale, its effect (even though partly masked by the more

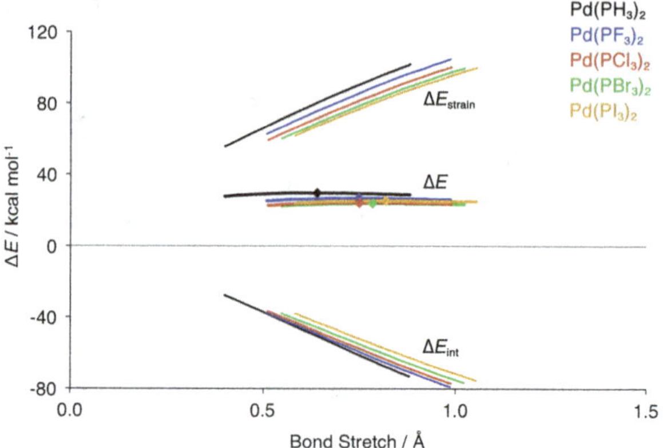

Fig. 10 Activation strain analyses [Eq. (1)] along partial energy profiles for Pd(PH$_3$)$_2$ and Pd(PX$_3$)$_2$, obtained by the TV-IRC method. Computed at dispersion-corrected ZORA-BLYP-D3/TZ2P

stabilizing electrostatic attraction and orbital interactions) is still clearly visible in the total interaction energy curve. Due to this direct ligand–ligand repulsion, the bite angle does not decrease any further from the TS to the PC of addition to Pd(PI$_3$)$_2$, but retains a value of 105.0°, slightly larger than for Pd(PBr$_3$)$_2$ (see Fig. 9).

Finally, we address the progressively weaker interaction between the fragments. From Pd(PH$_3$)$_2$ to the series of halogen-substituted catalysts, the interaction weakens mainly due to a less stabilizing orbital interaction term. This is caused by weaker catalyst-to-substrate backbonding, due to a lower orbital energy of the donating orbitals on the halogenated catalysts. The reason for these lower orbital energies is the better π-backdonation to the halogenated phosphine ligands (Table 1). This stronger backdonation generates a more positive potential on the Pd center, which stabilizes the donating orbitals. This is accompanied by the fact that, upon bending, the HOMO on the catalyst is pushed up less in energy, because from PH$_3$ to the halogen-substituted PX$_3$, the lone pairs are less localized on the phosphorus atom (see Sect. 4.2 and Fig. 8) and therefore have a weaker antibonding interaction with the palladium d orbitals.

Along the halogenated series, from Pd(PF$_3$)$_2$ to Pd(PI$_3$)$_2$, the orbital interaction term is remarkably similar, and the weakening of the catalyst–substrate interaction along this series results from an increasing Pauli repulsion. This destabilizing term is strengthened along this series because, from Pd(PF$_3$)$_2$ to Pd(PI$_3$)$_2$, there are more orbitals on the catalyst with energies in the vicinity of the methane HOMO energy, and therefore an increasingly large number of occupied catalyst orbitals enter a 2-center, 4-electron repulsion with the methane HOMO [87].

4.4 Conclusions and Outlook

Halogen-substituted palladium-phosphine complexes Pd(PX$_3$)$_2$ with X=F, Cl, Br, or I all have nonlinear geometries, unlike Pd(PH$_3$)$_2$ which has a linear ligand–metal–ligand angle. Along the Pd(PX$_3$)$_2$ series the ligand–metal–ligand angle decreases from 151.7° for Pd(PF$_3$)$_2$, to 143.2°, 136.6°, and 122.4° for Pd(PCl$_3$)$_2$, Pd(PBr$_3$)$_2$, and Pd(PI$_3$)$_2$, respectively. This follows from dispersion-corrected relativistic density functional computations. We found that the nonlinearity is the result of a combination of factors: firstly, the potential energy surfaces for bending the halogenated phosphine complexes are flat due to the increased π-backbonding that occurs upon bending from 180° to 90°. Secondly, from Pd(PH$_3$)$_2$ to Pd(PF$_3$)$_2$ and onwards to Pd(PI$_3$)$_2$, there is a less steeply increasing Pauli repulsion between PdPH$_3$ or PdPX$_3$ and the second ligand due to a smaller overlap of the highest occupied MOs upon bending. Thirdly, as along the Pd(PX$_3$)$_2$ series the halogen substituents become heavier, the stronger dispersion interactions between the ligands pull them more closely to each other.

When applied as catalytic complexes for methane C–H bond activation, this nonlinearity leads to a lower reaction barrier for the halogenated catalysts Pd(PX$_3$)$_2$ compared to Pd(PH$_3$)$_2$, because there is less deformation energy needed to bend away the ligands in order to make room for the approaching methane. Along the Pd(PX$_3$)$_2$ series, there are two opposing trends, resulting in lower barriers from Pd(PF$_3$)$_2$ to Pd(PBr$_3$)$_2$, but a slightly higher barrier for Pd(PI$_3$)$_2$. The two trends are: (1) a less destabilizing catalyst strain energy due to increased nonlinearity; counteracted by (2) a less stabilizing interaction energy due to a larger number of repulsive occupied–occupied orbital interactions. From Pd(PBr$_3$)$_2$ to Pd(PI$_3$)$_2$ this latter trend outweighs the effect of the decreased strain energy.

We envisage that the insights discussed in this chapter provide chemists with better design principles, and therefore contribute to a more rational fragment-oriented design of catalysts.

References

1. Diederich F, Stang PJ (1998) Metal-catalyzed cross-coupling reactions. Weinheim, Wiley-VCH
2. Hartwig JF (2010) Organotransition metal chemistry: from bonding to catalysis, 1st edn. University Science Books, Sausalito
3. The Nobel Prize in Chemistry (2010) Press Release. Nobelprize.org. http://nobelprize.org/nobel_prizes/chemistry/laureates/2010/press.html. 6 Oct 2010
4. Suzuki A (2011) Cross-coupling reactions of organoboranes: an easy way to construct C–C bonds (Nobel lecture). Angew Chem Int Ed 50:6722
5. Negishi E-I (2011) Magical power of transition metals: past, present, and future (Nobel lecture). Angew Chem Int Ed 50:6738
6. Shilov AE, Shul'pin GB (1997) Activation of C–H bonds by metal complexes. Chem Rev 97:2879
7. Beletskaya IP, Cheprakov AV (2000) The Heck reaction as a sharpening stone of palladium catalysis. Chem Rev 100:2009
8. Van Der Boom ME, Milstein D (2003) Cyclometalated phosphine-based pincer complexes: mechanistic insight in bond activation. Chem Rev 103:1759
9. Crabtree RH (2004) Organometallic alkane CH activation. J Organomet Chem 689:4083
10. Weisshaar JC (1993) Bare transition metal atoms in the gas phase: reactions of M, M$^+$, and M^{2+} with hydrocarbons. Acc Chem Res 26:213

11. Ziegler T (1991) Approximate density functional theory as a practical tool in molecular energetics and dynamics. Chem Rev 91:651

12. Niu SQ, Hall MB (2000) Theoretical studies on reactions of transition-metal complexes. Chem Rev 100:353

13. Torrent M, Solà M, Frenking G (2000) Theoretical studies of some transition-metal-mediated reactions of industrial and synthetic importance. Chem Rev 100:439

14. Dedieu A (2000) Theoretical studies in palladium and platinum molecular chemistry. Chem Rev 100:543

15. Diefenbach A, De Jong GT, Bickelhaupt FM (2005) Fragment-oriented design of catalysts based on the activation strain model. Mol Phys 103:995

16. Bickelhaupt FM (1999) Understanding reactivity with Kohn–Sham molecular orbital theory: E2-S_N2 mechanistic spectrum and other concepts. J Comp Chem 20:114

17. Diefenbach A, De Jong GT, Bickelhaupt FM (2005) Activation of H–H, C–H, C–C and C–Cl bonds by Pd and PdCl$^-$. Understanding anion assistance in C–X bond activation. J Chem Theory Comput 1:286

18. De Jong GT, Bickelhaupt FM (2007) Transition-state energy and position along the reaction coordinate in an extended activation strain model. ChemPhysChem 8:1170

19. Van Zeist W-J, Bickelhaupt FM (2010) The activation strain model of chemical reactivity. Org Biomol Chem 8:3118

20. Amatore C, Jutand A (1999) Mechanistic and kinetic studies of palladium catalytic sytems. J Organomet Chem 576:254

21. Kozuch S, Amatore C, Jutand A, Shaik S (2005) What makes for a good catalytic cycle? A theoretical study of the role of an anionic palladium(0) complex in the cross-coupling of an aryl halide with an anionic nucleophile. Organometallics 24:2319

22. Kozuch S, Shaik S (2011) How to conceptualize catalytic cycles? The energetic span model. Acc Chem Res 44:101

23. Kozuch S (2012) A refinement of everyday thinking: the energetic span model for kinetic assessment of catalytic cycles. WIREs Comput Mol Sci 2:795

24. Legault CY, Garcia Y, Merlic CA, Houk KN (2007) Origin of regioselectivity in palladium-catalyzed cross-coupling reactions of polyhalogenated heterocycles. J Am Chem Soc 129:12664

25. Galabov B, Nikolova V, Wilke JJ, Schaefer HF III, Allen WD (2008) Origin of the S_N2 benzylic effect. J Am Chem Soc 130:9887

26. Bento AP, Bickelhaupt FM (2008) Nucleophilicity and leaving-group ability in frontside and backside S_N2 reactions. J Org Chem 73:7290

27. Ess DH, Houk KN (2008) Theory of 1,3-dipolar. Cycloadditions: distortion/interaction and frontier molecular orbital models. J Am Chem Soc 130:10187

28. Wolters LP, Bickelhaupt FM (2012) Halogen bonding versus hydrogen bonding: a molecular orbital perspective. Chem Open 1:96

29. Bickelhaupt FM, Baerends EJ (2003) The case for steric repulsion causing the staggered conformation of ethane. Angew Chem Int Ed 42:4183

30. Poater J, Sola M, Bickelhaupt FM (2006) Hydrogen-hydrogen bonding in planar biphenyl, predicted by atoms-in-molecules theory, does not exist. Chem Eur J 12:2889

31. Fernández I, Bickelhaupt FM, Cossío FP (2012) Type-I dyotropic reactions: understanding trends in barriers. Chem Eur J 18:12395

32. Van Zeist W-J, Koers AH, Wolters LP, Bickelhaupt FM (2008) Reaction coordinates and the transition-vector approximation to the IRC. J Chem Theory Comput 4:920

33. Bickelhaupt FM, Baerends EJ (2000) Kohn–Sham density functional theory: predicting and understanding chemistry. In: Lipkowitz KB, Boyd DB (eds) Reviews in computational chemistry. Wiley-VCH, New York

34. Baerends EJ, Gritsenko OV (1997) A quantum chemical view of density functional theory. J Phys Chem A 101:5383

35. Ziegler T, Rauk A (1979) A theoretical study of the ethylene-metal bond in complexes between Cu$^+$, Ag$^+$, Au$^+$, Pt^0 or Pt^{2+} and ethylene, based on the Hartree–Fock–Slater transition-state method. Inorg Chem 18:1558
36. Te Velde G, Bickelhaupt FM, Baerends EJ, Fonseca GC, Van Gisbergen SJA, Snijders JG, Ziegler T (2001) Chemistry with ADF. J Comput Chem 22:931
37. Fonseca GC, Snijders JG, Te Velde G, Baerends EJ (1998) Towards an order-N DFT method. Theor Chem Acc 99:391
38. Baerends EJ, Ziegler T, Autschbach J, Bashford D, Bérces A, Bickelhaupt FM, Bo C, Boerrigter PM, Cavallo L, Chong DP, Deng L, Dickson RM, Ellis DE, Van Faassen M, Fan L, Fischer TH, Fonseca Guerra C, Ghysels A, Giammona A, Van Gisbergen SJA, Götz AW, Groeneveld JA, Gritsenko OV, Grüning M, Gusarov S, Harris FE, Van Den Hoek P, Jacob CR, Jacobson H, Jensen L, Kaminski JW, Van Kessel G, Kootstra F, Kovalenko A, Krykunov MV, Van Lenthe E, McCormack DA, Michalak A, Mitoraj M, Neugebauer J, Nicu VP, Noodleman L, Osinga VP, Patchkovskii S, Philipsen PHT, Post D, Pye CC, Ravenek W, Rodríguez JI, Ros P, Schipper PRT, Schreckenbach G, Seldenthuis JS, Seth M, Snijders JG, Solà M, Swart M, Swerhone D, Te Velde G, Vernooijs P, Versluis L, Visscher L, Visser O, Wang F, Wesolowski TA, Van Wezenbeek EM, Wiesenekker G, Wolff SK, Woo TK, Yakovlev AL, SCM, Theoretical Chemistry; Vrije Universiteit, Amsterdam, The Netherlands. http://www.scm.com/
39. Van Zeist W-J, Fonseca GC, Bickelhaupt FM (2008) PyFrag – streamlining your reaction path analysis. J Comput Chem 29:312
40. Diefenbach A, Bickelhaupt FM (2004) Activation of H–H, C–H, C–C, and C–Cl Bonds by Pd(0). Insight from the activation strain model. J Phys Chem A 108:8460
41. De Jong GT, Kovács A, Bickelhaupt FM (2006) Oxidative addition of hydrogen halides and dihalogens to Pd. Trends in reactivity and relativistic effects. J Phys Chem A 110:7943
42. Diefenbach A, Bickelhaupt FM (2005) Activation of C–H, C–C and C–I bonds by Pd and cis-Pd(CO)$_2$I$_2$. Catalyst–substrate adaptation. J Organomet Chem 690:2191
43. De Jong GT, Bickelhaupt FM (2007) Catalytic carbon–halogen bond activation: trends in reactivity, selectivity, and solvation. J Chem Theory Comput 3:514
44. Low JJ, Goddard WA III (1986) Theoretical studies of oxidative addition and reductive elimination. 2. Reductive coupling of H–H, H–C, and C–C bonds from palladium and platinum complexes. Organometallics 5:609
45. Low JJ, Goddard WA III (1986) Theoretical studies of oxidative addition and reductive elimination. 3. C–H and C–C reductive coupling from palladium and platinum bis(phosphine) complexes. J Am Chem Soc 108:6115
46. Blomberg MRA, Brandemark U, Siegbahn PEM (1983) Theoretical investigation of the elimination and addition reactions of methane and ethane with nickel. J Am Chem Soc 105:5557
47. Siegbahn PEM, Blomberg MRA (1992) Theoretical study of the activation of C–C bonds by transition metal atoms. J Am Chem Soc 114:10548
48. Saillard JY, Hoffmann R (1984) C–H and H–H activation in transition-metal complexes and on surfaces. J Am Chem Soc 106:2006
49. Forster D (1975) Halide catalysis of the oxidative addition of alkyl halides to rhodium (1) complexes. J Am Chem Soc 97:951
50. Blomberg MRA, Schüle J, Siegbahn PEM (1989) Ligand effects on metal-R bonding, where R is hydrogen or alkyl. A quantum chemical study. J Am Chem Soc 111:6156
51. Amatore C, Jutand A, Suarez A (1993) Intimate mechanism of oxidative addition to zerovalent palladium complexes in the presence of halide-ions and its relevance to the mechanism of palladium-catalyzed nucleophilic substitutions. J Am Chem Soc 115:9531
52. Siegbahn PEM, Blomberg MRA (1994) Halide ligand effects on the oxidative addition reaction of methane and hydrogen to second row transition metal complexes. Organometallics 13:354

53. Amatore C, Jutand A (2000) Anionic Pd(0) and Pd(II) intermediates in palladium-catalyzed heck and cross-coupling reactions. Acc Chem Res 33:314

54. Hammond GS (1955) A correlation of reaction rates. J Am Chem Soc 77:334

55. Fazaeli R, Ariafard A, Jamshidi S, Tabatabaie ES, Pishro KA (2007) Theoretical studies of the oxidative addition of PhBr to $Pd(PX_3)_2$ and $Pd(X_2PCH_2CH_2PX_2)$ (X=Me, H, Cl). J Organomet Chem 692:3984

56. Van Zeist W-J, Visser R, Bickelhaupt FM (2009) The steric nature of the bite angle. Chem Eur J 15:6112

57. Van Zeist W-J, Bickelhaupt FM (2011) Steric nature of the bite angle. A closer and a broader look. Dalton Trans 40:3028

58. Jover J, Fey N, Purdie M, Lloyd-Jones GC, Harvey JN (2010) A computational study of phosphine ligand effects in Suzuki–Miyaura coupling. J Mol Catal A Chem 324:39

59. Besora M, Gourlaouen C, Yates B, Maseras F (2011) Phosphine and solvent effects on oxidative addition of CH_3Br to $Pd(PR_3)$ and $Pd(PR_3)_2$ complexes. Dalton Trans 40:11089

60. Su M, Chu S (1998) Theoretical study of oxidative addition and reductive elimination of 14-electron d^{10} ML_2 complexes: a ML_2+CH_4 (M=Pd, Pt; L=CO, PH_3, $L_2=PH'_2CH_2CH_2PH_2$) case study. Inorg Chem 37:3400

61. Dierkes P, Van Leeuwen PWNM (1999) The bite angle makes the difference: a practical ligand parameter for diphosphine ligands. J Chem Soc Dalton 1999:1519–1529

62. Van Leeuwen PWNM, Kamer P, Reek JNH, Dierkes P (2000) Ligand bite angle effects in metal-catalyzed C–C bond formation. Chem Rev 100:2741

63. Freixa Z, Van Leeuwen PWNM (2003) Bite angle effects in diphosphine metal catalysts: steric or electronic? Dalton T 2003:1890

64. Birkholz (née Gensow) M-N, Freixa Z, Van Leeuwen PWNM (2009) Bite angle effects of diphosphines in C–C and C–X bond forming cross coupling reactions. Chem Soc Rev 38:1099

65. Hofmann P, Heiss H, Müller G (1987) Synthesis and molecular-structure of dichloro[η^2-bis (di-t-butylphosphino)methane]platinum(II), $Pt(dtbpm)Cl_2$. The electronic structure of 1,3-diphosphaplatinacyclobutane fragments. Z Naturforsch 42b:395

66. Koga N, Morokuma K (1991) Ab initio molecular orbital studies of catalytic elementary reactions and catalytic cycles of transition-metal complexes. Chem Rev 91:823

67. Eller K, Schwarz H (1991) Organometallic chemistry in the gas phase. Chem Rev 91:1121

68. Wittborn ACM, Costas M, Blomberg MRA, Siegbahn PEM (1997) The C–H activation reaction of methane for all transition metal atoms from the three transition rows. J Chem Phys 107:4318–4328

69. De Jong GT, Bickelhaupt FM (2009) Bond activation by group-11 transition-metal cations. Can J Chem 87:806

70. Wolters LP, Van Zeist W-J, Bickelhaupt FM (2014) d-regime, s-regime and intrinsic bite-angle flexibility: new concepts for designing d^{10}-ML_n catalysts. Submitted for publication

71. Wolters LP, Bickelhaupt FM (2013) Nonlinear d^{10}-ML_2 transition-metal complexes. Chem Open 2:106

72. Otsuka S (1980) Chemistry of platinum and palladium compounds of bulky phosphines. J Organomet Chem 200:191

73. Ziegler T (1985) Theoretical study on the stability of $M(PH_3)_2(O_2)$, $M(PH_3)_2(C_2H_2)$, and $M(PH_3)_2(C_2H_4)$ (M=Ni, Pd, Pt) and $M(PH_3)_4(O_2)^+$, $M(PH_3)_4(C_2H_2)^+$, and $M(PH_3)_4(C_2H_4)^+$ (M=Co, Rh, Ir) by the HFS-transition-state method. Inorg Chem 24:1547

74. King RB (2000) Atomic orbitals, symmetry, and coordination polyhedra. Coord Chem Rev 197:141

75. Carvajal MA, Novoa JJ, Alvarez S (2004) Choice of coordination number in d^{10} complexes of group 11 metals. J Am Chem Soc 126:1465

76. Albright TA, Burdett JK, Whangbo MH (2013) Orbital interactions in chemistry, 2nd edn. Wiley, New York

77. Zhou M, Andrews L (1998) Matrix infrared spectra and density functional calculations of $Ni(CO)_x^-$, $x=1$–3. J Am Chem Soc 120:11499

78. Manceron L, Alikhani ME (1999) Infrared spectrum and structure of Ni(CO)$_2$ – a matrix isolation and DFT study. Chem Phys 244:215
79. Jolly CA, Marynick DS (1989) Ground-state geometries and inversion barriers for simple complexes of early transition metals. Inorg Chem 28:2893
80. Kaupp M (2001) "Non-VSEPR" structures and bonding in d0 systems. Angew Chem Int Ed 40:3534
81. Gillespie RJ, Bytheway I, DeWitte RS, Bader RFW (1994) Trigonal bipyramidal and related molecules of the main group elements: investigation of apparent exceptions to the VSEPR model through the analysis of the laplacian of the electron density. Inorg Chem 33:2115
82. Kaupp M, Schleyer PVR (1992) The structural variations of monomeric alkaline earth MX$_2$ compounds (M=Ca, Sr, Ba; X = Li, BeH, BH$_2$, CH$_3$, NH$_2$, OH, F). An ab initio pseudopotential study. J Am Chem Soc 114:491
83. Kaupp M (1999) On the relation between π bonding, electronegativity, and bond angles in high-valent transition metal complexes. Chem Eur J 5:3631
84. Ogasawara M, Macgregor SA, Streib WE, Folting K, Eisenstein O, Caulton KG (1995) Isolable, unsaturated Ru(0) in Ru(CO)$_2$(P^{t}Bu$_2$Me)$_2$: not isostructural with Rh(I) in Rh(CO)$_2$(PR$_3$)$_2^+$. J Am Chem Soc 117:8869
85. Ogasawara M, Macgregor SA, Streib WE, Folting K, Eisenstein O, Caulton KG (1996) Characterization and reactivity of an unprecedented unsaturated zero-valent ruthenium species: isolable, yet highly reactive. J Am Chem Soc 118:10189
86. Grimme S, Antony J, Ehrlich S, Krieg H (2010) A consistent and accurate ab initio parametrization of density functional dispersion correction (DFT-D) for the 94 elements H-Pu. J Chem Phys 132:154104
87. Bickelhaupt FM, DeKock RL, Baerends EJ (2002) The short N–F bond in N$_2$F$^+$ and how Pauli repulsion influences bond lengths. Theoretical study of N$_2$X$^+$, NF$_3$X$^+$, and NH$_3$X$^+$ (X = F, H). J Am Chem Soc 124:1500

Struct Bond (2016) 167: 163–178
DOI: 10.1007/430_2015_178
© Springer International Publishing Switzerland 2015
Published online: 30 July 2015

The Electronics of CH Activation by Energy Decomposition Analysis: From Transition Metals to Main-Group Metals

Clinton R. King, Samantha J. Gustafson, and Daniel H. Ess

Abstract Alkane CH activation is a fundamental reaction class where a metal-ligand complex reacts with a CH bond to give a metal-alkyl organometallic intermediate. CH activation reactions have been reported for a variety of transition metals and main-group metals. This chapter highlights recent quantum-mechanical studies that have used energy decomposition analysis (EDA) to provide insight into σ-coordination complexes and transition states for alkane CH activation reactions. These studies have provided new conceptual understanding of CH activation reactions and detailed insight into the physical nature and magnitude of interaction between alkanes with transition metals and main-group metals.

Keywords CH activation · Energy decomposition analysis · Transition metals · Main-group metals

Contents

C.R. King, S.J. Gustafson, and D.H. Ess (✉)
Department of Chemistry and Biochemistry, Brigham Young University,
Provo, UT 84602, USA
e-mail: dhe@chem.byu.edu

Scheme 1 General CH bond functionalization strategy

CH Activation Rxn

$$MX + CH_4 \longrightarrow M{-}CH_3 + HX$$

MR Functionalization Rxn

$$M{-}CH_3 + 2HX + Ox \longrightarrow MX + CH_3X + H_2Ox$$

1 Introduction

CH activation reactions now play a dominant role in organometallic transformations and in synthetic organic reaction development [1–16]. In the context of this chapter, CH activation involves the reaction of a metal-ligand complex (MX) with an alkane CH bond to give an organometallic metal-alkyl (MR) intermediate that is hopefully more susceptible to substitution at carbon than the starting CH bond (Scheme 1) [17–20]. Generally, this CH bond functionalization strategy couples CH activation with a MR functionalization reaction where the organometallic intermediate is transformed into a partially oxidized alkane product. This two-step reaction sequence of CH activation and MR functionalization distinguishes this strategy from direct CH bond functionalization by metal oxo complexes or electrophilic organic reagents such as dioxiranes.

As experimental studies continue to discover new CH activation reactions, theory will play several significant roles such as (1) clarify and predict CH activation reaction mechanisms, (2) quantitatively predict CH activation transition-state barrier heights for new metal-ligand complexes, and (3) provide conceptual analysis of key CH activation intermediates and transition states. This last effort is the concern of this chapter and is important because conceptual analysis leads to understanding current reactivity and selectivity models and development of new models for the purpose of reaction discovery. Here we highlight select theoretical studies that have used energy decomposition analysis (EDA) methods to provide insight into intermediates and transition states for alkane CH activation reactions. Examples are discussed that provide detailed insight into the physical nature and magnitude of interaction of alkanes with transition metals and main-group metals. Nearly all examples discussed in this chapter have a direct relationship to experimental metal-ligand complexes that induce CH activation reactions.

2 Energy Decomposition Methods to Analyze CH Activation

There are several theoretical schemes to analyze the electronic structure of intermediates and transition states along a reaction pathway. Methods range from analysis of bonding using electron densities such as Bader's atoms-in-molecules

[21, 22] to second-order perturbation methods in natural bond orbital (NBO) analysis [23]. A particularly appealing method is EDA because this type of scheme relates the total energy of interaction (ΔE_{INT}) between two reacting molecules to mathematically well-defined and chemically meaningful types of interactions such as electrostatic energy, electron-electron repulsion energy, and electron delocalization energy. While there is no unique way to partition the energy of interaction between two molecules, the following provides a brief tutorial of how some of the most utilized EDA schemes dissect the interaction energy between reacting molecules. There are several excellent in-depth reviews of EDA methods and their application to chemical problems [24–31].

As two molecules react to form an intermediate or transition state, there are associated geometry changes and intramolecular and intermolecular electronic reorganization. For example, along the potential energy surface for CH activation of methane by a metal-ligand complex, there is CH bond stretching and alkane angle changes (Scheme 2). The metal-ligand complex also alters its geometry to accommodate the interaction with methane. This geometry change can be subtle if ligand rearrangement is unneeded but can be significant if, for example, a κ^2 ligand must transform into a κ^1 ligand. The energy associated with the difference in geometries between separated, noninteracting fragments and the reaction complex is called the distortion or activation-strain energy (ΔE_{DIST}, Eq. 1). The interaction between distorted fragments and the resulting energy stabilization is called the total interaction energy (ΔE_{INT}). The interaction energy between the distorted fragments controls the progress of geometric deformation required by the reactants to achieve the specific transition-state structure. The sum of the distortion and interaction energies gives the total energy of the bimolecular complex. For a transition-state structure, the sum of distortion and interaction energies is the activation energy:

Scheme 2 Relationship between distortion and interaction energies

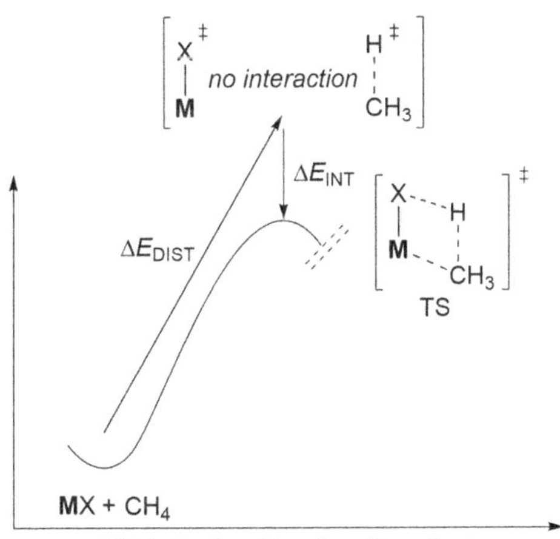

$$\Delta E = \Delta E_{\text{DIST}} + \Delta E_{\text{INT}} \tag{1}$$

While there are several insightful studies that analyze distortion energies [27, 29–31], this chapter emphasizes interaction energies and its dissection into specific physical terms. This dissection can be done in several different styles. Most EDA methods partition the interaction energy into components related to the classic frontier molecular orbital perturbation approach [32–36]. For example, the EDA method implemented in the Amsterdam Density Functional (ADF) program [24, 28, 37] partitions the interaction energy into contributions from electrostatic interactions (ΔE_{ELSTAT}), closed-shell or Pauli repulsion (ΔE_{PAULI}), and occupied-unoccupied orbital interactions (ΔE_{ORB}) (Eq. 2). Electrostatic interactions occur between nuclei-nuclei, nuclei-electron, and electron-electron. Closed-shell repulsion between interacting fragments arises from the requirement that same-spin electrons obey the Pauli exclusion principle. One way closed-shell repulsion is represented is by a classic 4-electron, 2-orbital interaction diagram shown on the left-hand side of Scheme 3. This diagram illustrates the net repulsion that occurs by having both bonding and antibonding orbitals filled. Delocalization orbital interactions result from the interaction of occupied and vacant orbitals. This is due to both intramolecular polarization and intermolecular interactions that describe electron delocalization. This latter type of occupied-vacant orbital interaction is described by the 2-electron, 2-orbital interaction diagram shown on the right-hand side of Scheme 3. This describes the energy stabilization resulting from partial electron density delocalization from the fragment with the occupied orbital to the fragment with the unoccupied orbital:

$$\Delta E_{\text{INT}} = \Delta E_{\text{ELSTAT}} + \Delta E_{\text{PAULI}} + \Delta E_{\text{ORB}} \tag{2}$$

While Eq. (2) represents the EDA method implemented in ADF, which is later referred to as ADF-EDA, there are several other implementations of EDA. One

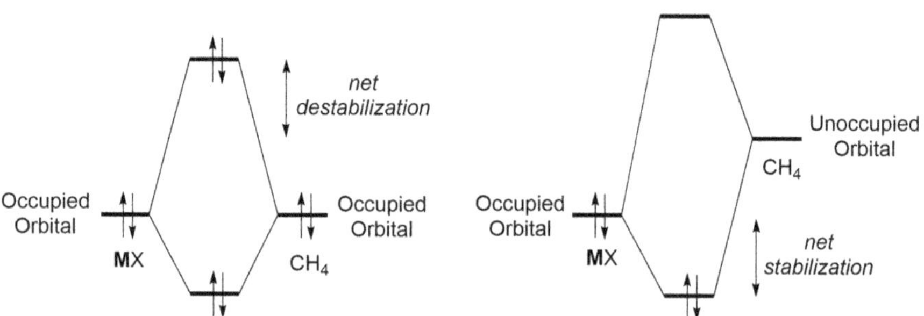

Scheme 3 Graphical illustration of Pauli repulsion (*left*) and occupied-unoccupied orbital stabilization upon charge transfer (*right*)

particularly useful method is the absolutely localized molecular orbital EDA (ALMO-EDA) [38, 39]. The ALMO-EDA method collapses the ΔE_{PAULI} and ΔE_{ELSTAT} terms into a single term that is called a frozen density term (ΔE_{FRZ}) Eq. (3). More importantly, the ALMO-EDA is a variational method that utilizes block localization of fragment molecular orbital coefficients to obtain directional intermolecular orbital stabilization, which is also labeled as charge transfer (ΔE_{CT}) stabilization (Eq. 4). Another advantage of the ALMO-EDA scheme is that this directional intermolecular orbital energy stabilization is separated from intramolecular orbital polarization stabilization (ΔE_{POL}). The orbital polarization stabilization is best related to intramolecular occupied-vacant orbital interactions that occur when one fragment is in the presence of the other fragment:

$$\Delta E_{INT} = \Delta E_{FRZ} + \Delta E_{POL} + \Delta E_{CT} \tag{3}$$

$$\Delta E_{CT} = \Delta E_{CT1} + \Delta E_{CT2} \tag{4}$$

The intermolecular orbital stabilization energy term, ΔE_{CT}, provides an estimate of all occupied-to-unoccupied orbital interactions between distorted fragments. However, the energy stabilization is often dominated by the interaction between the highest occupied and lowest unoccupied molecular orbitals (HOMO and LUMO) of each distorted fragment (Scheme 3). For CH activation reactions, the orbital interactions corresponding to electron delocalization from the metal-ligand complex to the alkane corresponds to back-bonding, ΔE_{CT1} (Scheme 4). The orbital interactions/electron delocalization from the alkane to the metal-ligand complex and the forward-bonding energy stabilization is ΔE_{CT2}. The sum of the back-bonding and forward-bonding intermolecular orbital interaction stabilization approximately gives the total intermolecular orbital stabilization energy in Eq. (4).

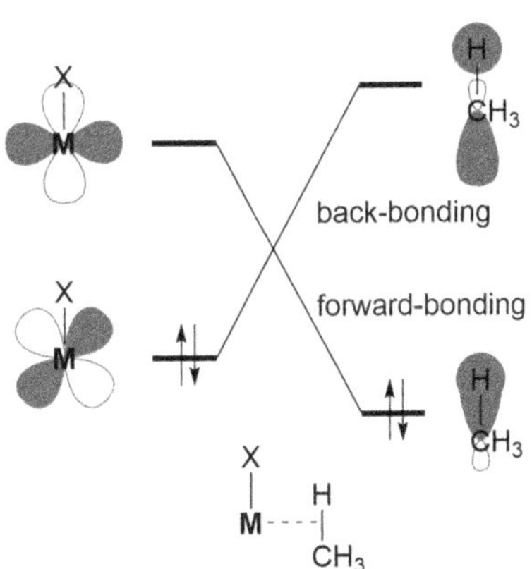

Scheme 4 Qualitative frontier orbital interactions that correspond to the back-bonding and forward-bonding electron delocalization

3   The Electronics of Alkane CH Bond Coordination

The CH activation reaction often occurs in two steps: (1) alkane coordination and (2) CH bond cleavage (Scheme 5). However, there are several examples of single-step reaction mechanisms. An alkane coordination complex (or σ-complex) involves the direct interaction of one or more alkane CH bonds with the inner sphere of the transition metal. There are now several examples of direct experimental detection of alkane complexes [40–42]. From a modeling perspective, an alkane coordination complex is an energy minimum structure on the potential energy landscape. Quantum-mechanical methods have been critical in characterizing these complexes and their energies, which can require accurate treatment of dispersion and spin-orbital coupling energies [43]. Alkane complexes are often exothermic (5–15 kcal/mol) relative to the starting metal-ligand complex and alkane if the metal has a vacant coordination site. However, alkane complexes are often endothermic relative to the starting metal-ligand complex and alkane if the metal is saturated with ligands and requires ligand dissociation prior to alkane inner-sphere contact with the metal center. Conceptually, the CH bond-metal interaction of an alkane complex involves both back-bonding and forward-bonding orbital interactions along with other intermolecular interactions. EDA provides a method to examine the chemical/physical driving force for interaction between the CH bond and the metal center. Historically, the stability of metal-alkane complexes has been attributed to back-bonding orbital interactions, and this is thought to be critical for strong interactions [1–16].

Bergman, Head-Gordon, and coworkers used the ALMO-EDA scheme to analyze the origin of alkane coordination to an experimentally relevant $CpRe(CO)_2$ complex [44]. As discussed earlier, the ALMO-EDA method is advantageous because it directly determines the relative amounts of forward-bonding and back-bonding orbital interactions in the context of their quantitative contribution to the total interaction energy. Scheme 6 shows how the EDA fragments were defined and

CH Activation Process

Step 1. Alkane Coordination · Step 2. CH Bond Cleavage

$$MX + CH_4 \xrightarrow{\text{Step 1. Alkane Coordination}} [M(CH_4)]^+ + X^- \xrightarrow{\text{Step 2. CH Bond Cleavage}} M-CH_3 + HX$$

Scheme 5 Two-step reaction sequence for CH activation

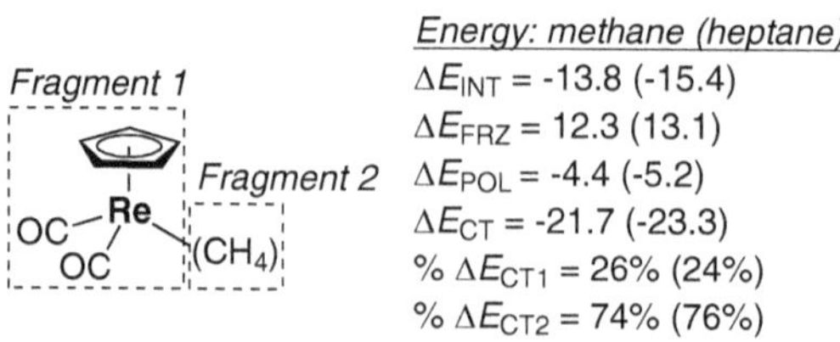

Scheme 6 BP86 ALMO-EDA energies reported by Cobar et al. [44] (kcal/mol)

the BP86 ALMO-EDA results reported by Bergman, Head-Gordon, and coworkers for the alkane complex interaction energies (ΔE_{INT}) of methane and heptane with $CpRe(CO)_2$. The total DFT interaction energies compare very well to experimental estimates.

The ALMO-EDA results in Scheme 6 reveal two interesting features of alkane coordination energy. First, the largest portion of the stabilizing interaction energy results from charge transfer as a result of orbital interactions. For example, the ΔE_{CT} energy stabilization for methane is -21.7 kcal/mol while ΔE_{POL} stabilization is only -4.4 kcal/mol. Second, in contrast to traditional concepts, the majority of alkane orbital stabilization energy (~75%) is the result of forward-bonding orbital interactions. Back-bonding orbital interactions are responsible for only ~25% of the orbital stabilization energy. This suggests that while the dominant interaction that stabilizes alkane complexes is the forward-bonding orbital interaction shown in Scheme 4, non-negligible back-bonding interactions also occur. The relative magnitude of back-bonding and forward-bonding orbital stabilization energies is reminiscent to that found in many metal-H_2 complexes.

Our group has also used the ALMO-EDA scheme to analyze methane complexes that are related to Ir, Ru, Os, and Rh metal-ligand complexes that are known to experimentally promote CH bond activation [45]. These metal-ligand complexes differ from $CpRe(CO)_2$ in that a ligand must be displaced prior to coordination. Therefore, EDA was performed on the alkane coordination complexes as well as the transition states en route to alkane coordination. Scheme 7 shows an example of the results reported for methane coordination to $(acac)_2Ir(OAc)$ via an associative coordination transition state. Similar to the results for the $CpRe(CO)_2(CH_4)$ complex, the energy stabilization for the $(acac)_2Ir(OAc)(CH_4)$ complex results mainly from forward-bonding orbital interactions. Despite the significant difference in ligand architecture, metal center, oxidation state, and endothermic alkane coordination, the forward-bonding interaction is responsible for ~80% of the orbital interaction stabilization energy while back-bonding is responsible for only ~20%

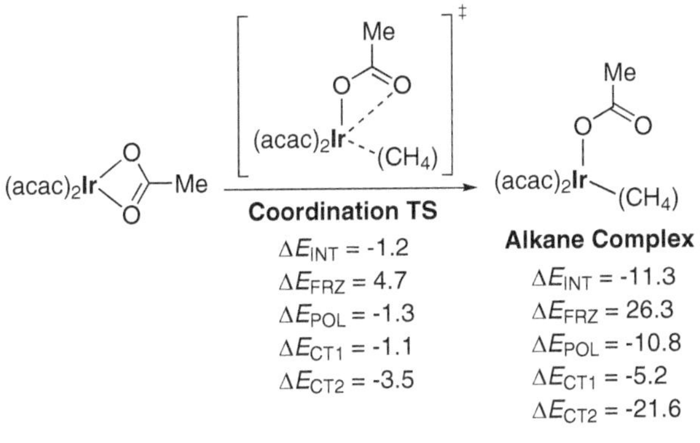

Scheme 7 B3LYP ALMO-EDA energies reported by Ess and coworkers [45]. EDA fragments were defined as $(acac)_2Ir(OAc)$ and CH_4 (acac = acetylacetonate, kcal/mol)

of orbital stabilization energy. However, inspection of the ALMO-EDA values for the coordination transition state reveals a very different description of the interaction of methane with the metal center and its ligands. Overall, there is only a very small interaction energy, and this suggests that the transition-state barrier for alkane coordination is dominated by acetate ligand dissociation to create a vacant coordination site for methane to approach and interact with the Ir metal center, and repulsion and stabilizing interactions are very weak between Ir and methane.

Another EDA study that was reported by our group focused on CH bond activation by Ir and Rh methyl, hydroxyl, and amido complexes with acac and Tp ligands [46]. For $(acac)_2IrX$, where $X = CH_3$ and OH, a methane coordination complex was identified on the potential energy surface. No minimum energy structure was located for the $(acac)_2IrNH_2$ complex. Similar results were found for the TpRu(CO)X complexes. Use of the ALMO-EDA method revealed that regardless of whether $X = CH_3$ or NH_2, the relative components to the interaction energy are similar. This report also highlighted that the ALMO-EDA scheme is relatively insensitive to the specific density functional used. B3LYP and M06 density functionals gave nearly identical results.

As an alternative to using EDA calculations to analyze back-bonding and forward-bonding orbital interactions, NBO-type calculations can also be used. The details of NBO calculations can be found elsewhere [23]. NBO analysis of donor-acceptor orbital interactions can be advantageous because the interpretation using localized orbitals and their interactions parallel the simplistic orbital interaction diagram shown in Scheme 4. However, the drawback of NBO calculations in the context of analyzing metal-alkane complexes is that estimates of back-bonding and forward-bonding orbital interactions are often not quantitatively directly related to the total interaction energy. Most of the time, the upside of NBO analysis outweighs its drawbacks and leads to useful conceptual understanding.

A recent example of NBO analysis of back-bonding and forward-bonding orbital interactions was performed on the $[(Cy_2PCH_2CH_2PCy_2)Rh(norbornane)]^+$ (Cy = cyclohexyl) alkane complex [47]. This alkane complex was directly detected by experiment in the solid state. Experiments and DFT calculations indicate that two CH bonds from one of the ethylene bridges of norbornane engage the Rh metal center. Similar to the ALMO-EDA results for the Re, Ir, and other alkane complexes, NBO analysis on $[(Cy_2PCH_2CH_2PCy_2)Rh(norbornane)]^+$ revealed that the major stabilizing interactions responsible for the stability of this alkane complex is forward-bonding interactions, which are approximately twice as stabilizing as back-bonding orbital interactions. In this instance, the key empty orbital on Rh is a Rh-P σ^* orbital that interacts with the filled CH σ orbitals. This interaction is twice as stabilizing as the sum of several filled Rh d_σ and Rh-P bonding orbitals interacting with the CH σ^* orbitals. Overall this and other NBO examples combined with the ALMO-EDA results suggest that for weakly interacting alkane complexes, the dominating stabilization is forward-bonding orbital interactions.

4 The Electronics of Alkane CH Bond Cleavage

Theory and experiment have identified a variety of CH bond cleavage mechanisms that result from either the alkane complex or directly from the metal-ligand complex and alkane [48, 49]. Because CH activation reactions have been discovered for many of the d-block transition metals, there has been a significant effort to classify and compare the mechanisms for CH bond cleavage. This is because often the specific choice of a metal and ligand results in a particular mechanism. Scheme 8 portrays oxidative addition and σ-bond metathesis mechanisms and their corresponding CH bond cleavage transition states, which represent the extremes of the commonly reported mechanisms. Intermediate between these extremes is a composite of the two transition states where the end result is a σ-bond metathesis mechanisms, but the metal is highly involved in bonding to the hydrogen in the transition state. This type of composite transition state has several names for nuanced versions, such as oxidatively added transition state (OATS) [50, 51], metal-assisted σ-bond metathesis (MAσBM) [52, 53], or σ-complex assisted metathesis (σ-CAM) [54]. Hall and Vastine [55] have shown that using atoms-in-molecules analysis provides a framework to understand the entire set of transition states along this mechanistic continuum.

Rather than focus on classifying CH activation reactions by mechanisms, our lab was interested in understanding and classifying the electronic reorganization that occurs in CH bond cleavage transition states as a possible tool to elucidate design principles for new CH activation reactions. Similar to the analysis of alkane coordination complexes, we used the EDA approach to probe the components of the interaction energy in the transition states. One potential advantage of an electronic reorganization viewpoint for classifying CH activation reactions is that many different metals and ligands can be analyzed without consideration of mechanism labels. Broadly, we considered two mechanisms labeled as insertion and substitution. Insertion transition states lead to a metal-hydride intermediate while substitution transition states pass the alkane hydrogen to a ligand. Scheme 9 shows insertion and substitution transition states and how EDA fragmentation was assigned.

In discussing electronic reorganization in the transition state, we have adopted the classic terms of electrophile and nucleophile that were previously used by Ryabov [56] for CH activation transition states. In the context of CH activation reactions, these terms are metal centric and refer to the overall influence of the

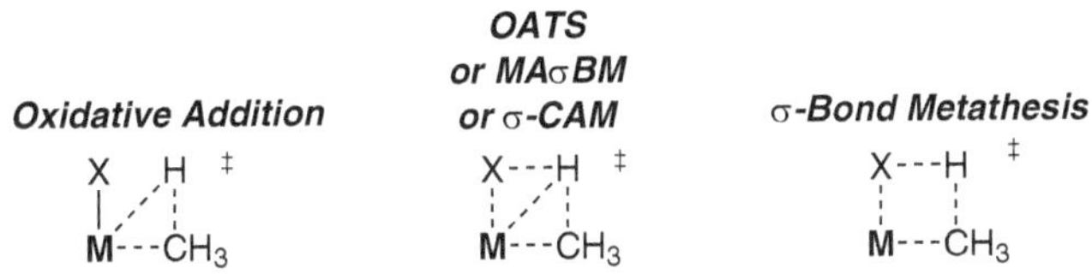

Scheme 8 Examples of mechanisms identified by theory and experiment for CH bond cleavage

                                                                C.R. King et al.

Scheme 9 Definition of EDA fragments for insertion and substitution transition states

metal-ligand fragment on the alkane. While there is no unique way to measure electrophilicity or nucleophilicity in a transition state, one useful approach is to estimate the energy stabilization resulting from electron delocalization due to orbital interactions. Outlined in Scheme 4, the simplified frontier orbital interactions that occur for alkane complexes also occur in CH bond cleavage transition states but are amplified. An electrophilic transition state is defined by a transition state where the majority of energy stabilization that occurs upon interaction between the metal-ligand and alkane results from forward-bonding orbital interactions. A nucleophilic transition state is defined by a transition state where the majority of energy stabilization that occurs upon interaction between the metal-ligand and alkane results from back-bonding orbital interactions. Whether back-bonding or forward-bonding orbital interactions dominate in the transition state is related to relative orbital energies of the metal-ligand complex and alkane and orbital overlap.

The ALMO-EDA scheme was used to analyze a wide range of methane CH bond cleavage transition states for metal-ligand complexes that have been reported experimentally to promote or catalyze CH activation [57]. This included late transition metals such as Rh(III), Rh(I), Ir(III), Pt(II), Pd(II), Ru(II), and Au(III) and early transition metals such as Sc(III) and W(II). Ligands attached to these metal centers range from phosphine, N-heterocycle, and halide to O-donors. Three specific examples of transition states analyzed and their ALMO-EDA interaction energy components are shown in Scheme 10.

This ALMO energy analysis revealed several important features about the nature of CH bond cleavage transition states: (1) By plotting the difference and ratio of ΔE_{CT1} and ΔE_{CT2} for the transition states, this revealed that there is a continuum from electrophilic to nucleophilic with several metal-ligand complexes that act as ambiphiles towards methane. The ΔE_{FRZ} and ΔE_{POL} energy terms were not considered in electrophilic/nucleophilic character assignment since these terms do not result from intermolecular electron delocalization. This continuum is illustrated in Scheme 10 by [(Hbpym)PtCl]$^{+2}$ that induces a transition state with methane where the dominant orbital stabilization energy results from forward-bonding (ΔE_{CT2}) and on the opposite extreme (PCP)Ir that induces a transition state with methane where the dominant orbital stabilization energy is back-bonding (ΔE_{CT1}). (2) Insertion and substitution-type mechanisms were found for electrophilic, ambiphilic, and nucleophilic transition states. (3) Terms such as oxidative addition do not imply electronic changes in the transition state. For example, the label oxidative addition is often used for insertion reactions between Pt(II) and alkanes. However, for

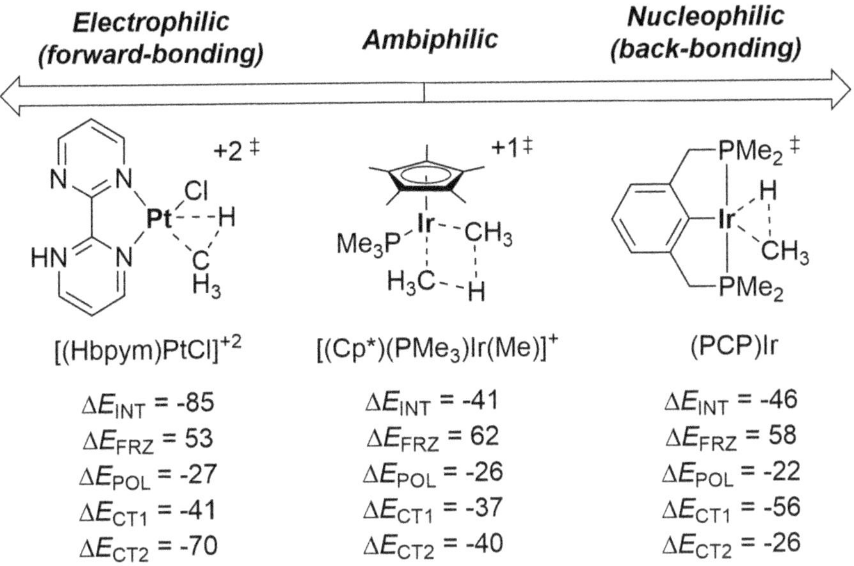

Scheme 10 Examples of electrophilic, ambiphilic, and nucleophilic metal-ligand complexes [57] (kcal/mol)

$[(Hbpym)PtCl]^{+2}$ reacting with methane, the oxidative addition transition state is electrophilic and dominated by donation of electron density from methane to Pt. In contrast, the (PCP)Ir complex induces oxidative addition with the main orbital energy stabilization to the transition state interaction energy involving electron density delocalization from (PCP)Ir to methane. (4) Although there was no quantitative correlation between the relative ALMO-EDA charge transfer energy stabilization and frontier orbital energy gaps, transition-state fragment orbital energies were found to qualitatively reflect electrophilic, ambiphilic, or nucleophilic character.

Overall, late transition metals, such as Au(III), Rh(III), Pd(II), and Pt(II), with a variety of N-donor and O-donor ligands were often found to have electrophilic transition states. Ru(II) and Ir(III) with N-donor and O-donor ligands as well as $[Cp*(PMe_3)IrMe]^+$ were found to act in an ambiphilic manner towards methane. Despite substitution-type transition states having both a Lewis acid metal and a Lewis base ligand that interacts with the CH bond of methane, these transition states are not always ambiphilic. For example, in the transition state between d^0 $(Cp*)_2ScMe$ and methane, the $(Cp*)_2ScMe$ complex acts as a nucleophile. Other nucleophilic transition states involved methane reaction with W(II), Ir(I), and Rh (I) complexes.

An ab initio molecular dynamics (AIMD) study by Vidossich, Ujaque, and Lledós examined the reaction pathway for methane reacting with $Cl_2Pd(H_2O)$ [58]. This study verified that this and similar Pt(II) complexes act as electrophiles towards methane. Along the reaction pathway and at the $Cl_2Pd(H_2O)(H)$

Scheme 11 Comparison of ALMO-EDA and ADF-EDA results for the CH activation cleavage transition state between $CsNH_2$ and methane (kcal/mol)

Substitution TS

$H_2N\text{-}\text{-}\text{-}H \quad ^{\ddagger}$

$\overset{|}{Cs}\text{-}\text{-}\overset{|}{CH_3}$

B3LYP	**BLYP**
ALMO-EDA	**ADF-EDA**
$\Delta E_{INT} = -26.5$	$\Delta E_{INT} = -25.4$
$\Delta E_{FRZ} = 73.7$	$\Delta E_{PAULI} = 147.2$
$\Delta E_{POL} = -48.7$	$\Delta E_{ESTAT} = -77.9$
$\Delta E_{CT1} = -49.2$	$\Delta E_{ORB} = -94.7$
$\Delta E_{CT2} = -2.3$	

(Me) intermediate, maximally localized Wannier functions were used to analyze the oxidation state of Pt. These localized bonding functions indicate that the Pt metal center remains Pt(II) throughout the reaction pathway without significant oxidation, which is in line with a highly electrophilic CH activation process.

To further examine the nucleophilic side of the CH activation continuum, the our group and the Cundari group undertook a computational study [59] of alkali metal amide and alkaline earth metal amide CH activation reactions with alkanes. This study is directly related to classic superbase chemistry using cesium cyclohexylamide-type reagents to induce CH bond deprotonation of alkanes [60].

While simple model systems using a single metal amide complex were used, this study illustrates a few important points. Scheme 11 shows the results reported using ALMO-EDA and ADF-EDA to analyze the transition state between $CsNH_2$ and methane. Importantly, while these different EDA schemes have a high degree of compatibility, they provide different information. For example, the total interaction energies are within 1 kcal/mol, and the ΔE_{FRZ} term is very close to the sum of the ΔE_{ELSTAT} and ΔE_{PAULI} ADF-EDA terms. The B3LYP ΔE_{FRZ} energy is 73.7 kcal/mol while the sum of the BLYP ΔE_{ELSTAT} and ΔE_{PAULI} terms is 69.3 kcal/mol.

The B3LYP ALMO-EDA values confirm the expected highly nucleophilic character of this transition state, which is the result of the amide group interacting with the methane CH σ^* and results in -49.2 kcal/mol of stabilization energy. Forward-bonding due to the interaction between the methane σ CH bond and the Cs vacant orbital is only -2.3 kcal/mol stabilizing.

For highly nucleophilic (or electrophilic) transition states, it is important to also examine the interaction energy components where electrostatic interactions are separated from closed-shell (Pauli) repulsion energy. The ADF-EDA interaction energy values in Scheme 11 indicate that in the $CsNH_2$-methane transition state the interaction energy has nearly equal energy stabilization from orbital interactions and electrostatic interactions. This is important because highly polarized transition states are often assumed to have very little covalent (i.e., orbital) energy stabilization. The ALMO-EDA energy components also reveal that half of this orbital stabilization is due to intramolecular orbital polarization (ΔE_{POL}) and half is due to electron delocalization (ΔE_{CT}). This illustrates that in general very polar transition states with significant electrostatic stabilization will induce a large amount of

Scheme 12 Comparison of ADF-EDA results reported by Bickelhaupt and coworkers for methane insertion by Pd(0) and Ca(0) [61–64] (kcal/mol)

Insertion TS

$$Pd\overset{\ddagger}{\underset{CH_3}{\cdots H}} \qquad Ca\overset{\ddagger}{\underset{CH_3}{\cdots H}}$$

BLYP ADF-EDA

$\Delta E_{INT} = -48.3$
$\Delta E_{PAULI} = 201.7$
$\Delta E_{ESTAT} = -159.1$
$\Delta E_{ORB} = -90.9$

BLYP ADF-EDA

$\Delta E_{INT} = -23.0$
$\Delta E_{PAULI} = 164.9$
$\Delta E_{ESTAT} = -83.5$
$\Delta E_{ORB} = -104.4$

intramolecular polarization energy stabilization. Overall, this study showed that the combination of using the ALMO-EDA and ADF-EDA methods provides a unified and insightful view of highly polarized CH activation transition states.

The Bickelhaupt group has applied the ADF-EDA scheme to a number of model alkane CH activation reactions (for select examples, see [61–64]). In one report, insertion transition states were analyzed for model CH activation reactions between methane and zero-valent group 2 (Be, Mg, and Ca) and group 12 (Zn and Cd) metals [61–64]. One motivation of this study was to provide an understanding based on transition state interactions why the lowest oxidation state of s-block main-group elements compared to Pd(0) have dramatically larger activation barriers for methane CH bond insertion. However, more relevant to this discussion is that the ADF-EDA given in Scheme 12 provide insight into the differences in transition-state electronic reorganization for s-block versus transition-metal elements for CH bond cleavage of methane. Surprisingly, in the Ca transition state, there is a nearly equal amount of electrostatic and orbital stabilization, while in the Pd transition state the electrostatics dominate the stabilizing interaction energy.

The Bickelhaupt group also examined model CH activation reactions between methane and low-oxidation state group-11 transition-metal cations Cu(I), Ag(I), and Au(I), with a comparison to Pd(0) [61–64]. This study revealed that at the beginning of the reaction pathway the low-oxidation state group II cations are more effective electrophiles than Pd(0) due to the much lower energy of the 4s, 5s, or 6s LUMOs. However, as the reaction progresses towards the transition state, back-bonding interactions become more important, and the group II cations are less effective at back-bonding compared to Pd(0).

Recently, the Periana group disclosed that main-group Tl(III) and Pb(IV) trifluoroacetate metal-ligand complexes promote alkane partial oxidation, and our group showed that this occurs via a CH activation mechanism (Scheme 13) [65]. While the above discussion has highlighted insights about the electronic character of CH activation reactions for transition-metal complexes, there is relatively little known about the electronic character for CH activation by p-block main-group metals. The ALMO-EDA and ADF-EDA component energies for the rate-limiting transition state between Tl(TFA)$_3$ and methane are reported in Scheme 13.

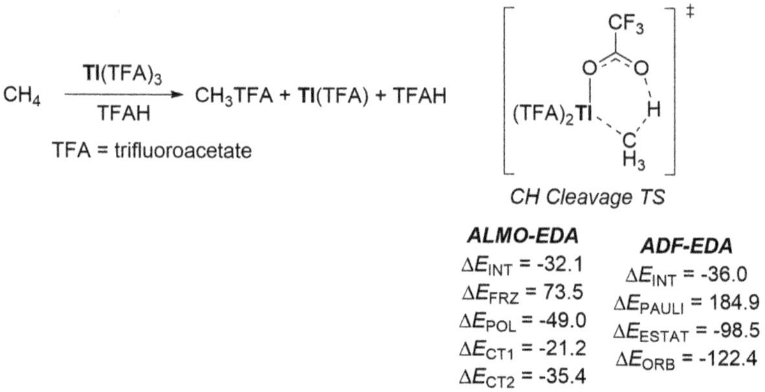

$$CH_4 \xrightarrow[\text{TFAH}]{\text{Tl(TFA)}_3} CH_3TFA + Tl(TFA) + TFAH$$

TFA = trifluoroacetate

CH Cleavage TS

ALMO-EDA

$\Delta E_{INT} = -32.1$

$\Delta E_{FRZ} = 73.5$

$\Delta E_{POL} = -49.0$

$\Delta E_{CT1} = -21.2$

$\Delta E_{CT2} = -35.4$

ADF-EDA

$\Delta E_{INT} = -36.0$

$\Delta E_{PAULI} = 184.9$

$\Delta E_{ESTAT} = -98.5$

$\Delta E_{ORB} = -122.4$

Scheme 13 Tl(III)-promoted methane CH activation transition state and ALMO-EDA (M06/6-31+G(d,p)[LANL2DZ]) and ADF-EDA (BP86/TZ2P-ZORA) component energies (kcal/mol)

Comparison of ΔE_{CT1} and ΔE_{CT2} indicate, as expected, that Tl(III) acts as a relatively strong electrophile towards methane with the main orbital stabilization energy resulting from forward-bonding orbital interactions in the transition state. While orbital interactions provide the largest ALMO-EDA stabilization energy to the total interaction energy, polarization stabilization energy is nearly as significant in magnitude. Because the polarization energy stabilization is large, it is important to also examine the ADF-EDA values that analyze electrostatic stabilization. The ADF-EDA values reported in Scheme 13 indicate that orbital energy stabilization is larger than electrostatic stabilization, but when polarization is considered from the ALMO-EDA the transition state is nearly equally comprised of stabilization from charge transfer and electrostatic interactions.

5 Conclusions

The theoretical studies highlighted here used EDA to provide detailed insights into the physical interactions that occur along the reaction pathway for CH activation. Use of the ALMO-EDA scheme revealed that weak alkane complexes form primarily as a result of forward-bonding orbital interactions. In contrast, transition states that involve CH bond cleavage can either be dominated by forward-bonding or back-bonding interactions, which are, respectively, referred to as electrophilic and nucleophilic transition states. There are also transition states that have nearly equal energy stabilization from forward-bonding or back-bonding interactions. While many details are known about the electronic reorganization during CH activation by transition metals, there is relatively little known about CH activation reactions with main-group metals. EDA applied to the Tl(III)-methane CH activation transition state revealed that Tl(III) reacts as a strong electrophile and that there is nearly equal energy stabilization from orbital interactions and electrostatic interactions.

References

1. Crabtree RH (1985) Chem Rev 85:245
2. Shilov AE, Shul'pin GB (1987) Russ Chem Rev 56:442
3. Arndtsen BA, Bergman RG (1995) Science 270:1970
4. Crabtree RH (1995) Chem Rev 95:987
5. Shilov AE, Shul'pin GB (1997) Chem Rev 97:2879
6. Stahl SS, Labinger JA, Bercaw JE (1998) Angew Chem Int Ed 37:2180
7. Crabtree RH (2001) J Chem Soc Dalton Trans 2437
8. Labinger JA, Bercaw JE (2002) Nature 417:507
9. Ritleng V, Sirlin C, Pfeffer M (2002) Chem Rev 102:1731
10. Jones WD (2003) Acc Chem Res 36:140
11. Goldman AS, Goldberg KI (2004) Organometallic C–H bond activation: an introduction. In: Goldman AS, Goldberg KI (eds) Activation and functionalization of C–H bonds, vol 885, ACS symposium series. Wiley, Washington, p 1
12. Crabtree RH (2004) J Organomet Chem 689:4083
13. Labinger JA (2004) J Mol Catal A Chem 220:27
14. Hashiguchi BG, Hövelmann CH, Bischof SM, Lokare KS, Leung CH, Periana RA (2010) Methane-to-methanol conversion. In: Crabtree RH (ed) Energy production and storage: inorganic chemical strategies for a warming world, Encyclopedia of inorganic chemistry. Wiley, Chichester, p 101
15. Gunnoe TB (2012) In: Perez PJ (ed) Alkane C–H activation by single-site metal catalysts, vol. 38. Springer, Dordrecht, pp 1–15
16. Cavaliere VN, Wicker BF, Mindiola DJ (2012) Adv Organomet Chem 60:1
17. Conley BL, Tenn WJ III, Young KJH, Ganesh SK, Meier SK, Ziatdinov VR, Mironov O, Oxgaard J, Gonzales J, Goddard WA III, Periana RA (2006) J Mol Catal A 251:8
18. Webb JR, Bolaño T, Gunnoe TB (2011) ChemSusChem 4:37
19. Golisz SR, Gunnoe TB, Goddard WA III, Groves JR, Periana RA (2011) Catal Lett 141:213
20. Hashiguchi BG, Bischof SM, Konnick MM, Periana RA (2012) Acc Chem Res 45:885
21. Bader R (1990) Atoms in molecules: a quantum theory. Oxford University Press, New York
22. Popelier PLA (2014) The QTAIM perspective of chemical bonding. In The chemical bond. Wiley-VCH , Weinheim, pp 271–308
23. Weinhold F, Landis CR (2012) Discovering chemistry with natural bond orbitals. Wiley, Hoboken
24. Bickelhaupt FM, Baerends EJ (2000) Kohn-sham DFT: predicting and understanding chemistry. In: Boyd DB, Lipkowitz KB (eds) Reviews in computational chemistry, vol 15. Wiley-VCH, New York, pp 1–86
25. von Hopffgarten M, Frenking G (2012) WIREs Comput Mol Sci 2:43
26. Lein M, Frenking G (2005) The nature of the chemical bond in the light of an energy decomposition analysis. In: Dykstra CE, Frenking G, Kim KS, Scuseria GE (eds) Theory and applications of computational chemistry: the first forty years. Elsevier , Amsterdam, pp 291–372
27. van Zeist WJ, Bickelhaupt FM (2010) Org Biomol Chem 8:3118
28. Frenking G, Bickelhaupt FM (2014) The EDA perspective of chemical bonding. In: Frenking G, Shaik S (eds) The chemical bond. Wiley-VCH, Weinheim, pp 121–157
29. Fernández I (2014) Phys Chem Chem Phys 16:7662
30. Fernández I, Bickelhaupt FM (2014) Chem Soc Rev 43:4953
31. Fernández I (2014) Understanding trends in reaction barriers. In: Pignataro B (ed) Discovering the future of molecular sciences. Wiley-VCH, Weinheim, pp 165–187
32. Klopman G (1968) J Am Chem Soc 90:223
33. Salem L (1968) J Am Chem Soc 90:543
34. Salem L (1968) J Am Chem Soc 90:553
35. Morokuma K (1971) J Chem Phys 55:1236

36. Ziegler T, Rauk A (1977) Theor Chim Acta 46:1
37. ADF (2014) SCM theoretical chemistry. Vrije Universiteit, Amsterdam. http://www.scm.com
38. Khaliullin RZ, Cobar EA, Lochan RC, Bell AT, Head-Gordon M (2007) J Phys Chem A 111:8753
39. Khaliullin RZ, Bell AT, Head-Gordon M (2008) J Chem Phys 128:184112
40. Walter MD, White PS, Schauer CK, Brookhart M (2013) J Am Chem Soc 135:15933
41. Pike SD, Thompson AL, Algarra AG, Apperley DC, Macgregor SA, Weller AS (2012) Science 337:1648
42. Bernskoetter WH, Schauer CK, Goldberg KI, Brookhart M (2009) Science 326:553
43. Chan B, Ball GE (2013) J Chem Theory Comput 9:2199
44. Cobar EA, Khaliullin RZ, Bergman RG, Head-Gordon M (2007) Proc Nat Acad Sci USA 104:6963
45. Ess DH, Bischof SM, Oxgaard J, Periana RA, Goddard WA III (2008) Organometallics 27:6440
46. Ess DH, Gunnoe TB, Cundari TR, Goddard WA III, Periana RA (2010) Organometallics 29:6801
47. Pike SD, Chadwick FM, Rees NH, Scott MP, Weller AS, Kramer T, Macgregor SA (2015) J Am Chem Soc 137:820
48. Balcells D, Clot E, Eisenstein O (2010) Chem Rev 110:749
49. Ackermann L (2011) Chem Rev 111:1315
50. Ng SM, Lam WH, Mak CC, Tsang CW, Jia G, Lin Z, Lau CP (2003) Organometallics 22:641
51. Lam WH, Jia G, Lin Z, Lau CP, Eisenstein O (2003) Chem Eur J 9:2775
52. Webster CE, Fan Y, Hall MB, Kunz D, Hartwig JF (2003) J Am Chem Soc 125:858
53. Hartwig JF, Cook KS, Hapke M, Incarvito CD, Fan Y, Webster CE, Hall MB (2005) J Am Chem Soc 127:2538
54. Perutz RN, Sabo-Etienne S (2007) Angew Chem Int Ed 46:2578
55. Vastine BA, Hall MB (2007) J Am Chem Soc 129:12068
56. Ryabov AD (1990) Chem Rev 90:403
57. Ess DH, Goddard WA III, Periana RA (2010) Organometallics 29:6459
58. Vidossich P, Ujaque G, Lledós A (2012) Chem Commun 48:1979
59. Pardue DB, Gustafson SJ, Periana RA, Ess DH, Cundari TR (2013) Comput Theor Chem 1019:85
60. Streitwieser A Jr, Ciuffarin E, Hammons JH (1967) J Am Chem Soc 89:63
61. Diefenbach A, de Jong GT, Bickelhaupt FM (2005) J Chem Theory Comput 1:286
62. Wolters LP, van Zeist WJ, Bickelhaupt FM (2014) Chem Eur J 20:11370
63. de Jong GT, Visser R, Bickelhaupt FM (2006) J Organomet Chem 691:4341
64. de Jong GT, Bickelhaupt FM (2009) Can J Chem 87:806
65. Hashiguchi BG, Konnick MM, Bischof SM, Gustafson SJ, Devarajan D, Gunsalus N, Ess DH, Periana RA (2014) Science 343:1232

Index

CPSIA information can be obtained
at www.ICGtesting.com
Printed in the USA
LVHW081554300619
622783LV00003B/131/P